静 谧 与 欢 悦
TRANQUILITY & JOY

张 唐 景 观　Z + T STUDIO：2009 — 2018

上海张唐景观设计事务所　著

同济大学出版社
Tongji University Press
中国·上海

序：寻找衍生品时代的真实

张海　摄影师，视觉艺术家，常居美国纽约

2008 年，整个世界都在经历巨大的变化。对于我而言，就是告别建筑师的职业，开始了一个艺术家的冒险生涯。有一天，张东和子颖从波士顿打来电话，说他们打算回国开始自己的事业。这让我非常惊讶。难以想象热爱自然、性情低调、有些书生气，同时对审美又固执己见的这对夫妇如何在中国生存，更不要说经营自己的公司。

2009 年，在一项研究基金的资助下，我开始频繁往返于定居的美国纽约和中国之间拍摄照片。张东和子颖在上海创办了他们的小工作室，并且雇了几个帮手。每次途经上海，我都会到他们的新家一起吃顿晚饭。每次吃饭的时候，他们除了抱怨经营公司不容易和上海冬天的难熬，并没有提及太多他们设计的东西。

一年后，也就是 2010 年，在我出发去中国之前，他们联系我问可不可以给他们上海的项目摄影。项目是一个精品会所的庭院景观，规模比较小。虽然我回复说当然可以，心里却不知道面对的会是什么。在 10 月下旬的一个下午，我抵达上海机场，然后就直奔项目现场。会所重重的铁门慢慢打开，我对眼前看到的景象激动不已——这个庭院不仅和周围被过度装饰的殖民时期的建筑形成鲜明对比，还将整个空间转化成了一个现代的、极简的、充满活力的场所。

在接下来的近十年里，我有幸为张唐景观大部分新完成的项目拍摄了照片。通过摄影，我们记录和调研了几乎所有张东和子颖设计的项目，甚至还包括他们去美国之前做的。我看到一些他们以前从来没有说过的东西。

迄今为止，我在中国的摄影足迹已走过十年。我目睹了中国从改革开放到现在 40 年来最有当代性的时刻：一个国家从关注如何提高生产力到如何把所有的东西都商品化；整个社会都在经历着充满不确定性的巨大压力，在过去和未来之间苦苦挣扎。事物本身的价值受到质疑，大家更关注的是其附加值。对于诚实创作的设计师来说，这种境况何其艰难。在这样的社会里，我们不得不问一个不可回避的核心问题——什么是真实的？

　　在不到十年的时间里,张东和子颖不仅在中国的景观行业里获得了充分肯定,而且两次获得美国景观学会的重要奖项,得到了世界范围的认可。他们的设计有多种尺度、多种类型,涵盖了不同材料,涉及不同技术。对于大多数的人来说,这样的成绩当然是成功,并且令人羡慕。但是,我始终不能承认我对他们当年要回国发展的担心是完全错误的。对于我来讲,我认识的他们始终没有改变——那对冬天坐在大西洋岸边的一辆旧车里安静地画水彩的夫妇。从他们的项目里,我始终不能把张东和子颖与当代的中国语境联系在一起。在我的脑海里,这两者总是有些错位。毋庸置疑地讲,中国特殊的环境是促成他们成功的重要原因之一,虽然环境可以使然,然而只有设计师本身才可以让事物成为真实的。

　　我们生活在一个衍生品的时代,但是我相信,张东和子颖并不是为了自己而追求真实,因为他们的作品足以代表真实。

张海

2018年2月14日,纽约

Foreword : Looking for the Truth in an Era of Derivatives

Hai Zhang——photographer and visual artist based in New York City

There were quite a few big changes around the world in the year 2008. I left my job as an architect to begin an adventure as an artist. One day, Dong and Ziying called me from Boston saying that they were planning to move back to China to start their own business. It caught me in surprise. I just could not imagine how this couple who are shy, humble, bookish, deeply in love with nature, yet stubborn in terms of aesthetic could survive in China, not to mention run their own business.

In 2009, with a research grant I begun frequently to go back and forth between my resided New York City and China to take photographs; Dong and Ziying started their own little studio in Shanghai and had hired a few helpers. Whenever I was there, I always stopped by their new home for dinner. They complained that the business was not easy and that the winter in Shanghai was depressing. During the whole conversation, however, they did not mention much about what they were designing.

A year later, in 2010, before one of my trips back to China, they contacted me and asked if I would be willing to photograph one of their projects in Shanghai. They said it was a small piece for a courtyard in an elite club. I agreed, not knowing what to expect. On a late October afternoon, I arrived at Shanghai airport and went straight to the project. When the heavy metal gate of the club opened, I was thrilled by what I saw —— a courtyard that not only drew aesthetic contrast to the surrounding overdecorated colonial style mansions, but also transformed the entire space to a modern, simple, yet playful place.

For the following near-decade, I have had privilege to photograph Z+T's newly completed projects while conducting a photographic survey of almost all the projects designed by Dong and Ziying, including many they completed before heading to the US. I had a chance to see all the works that they didn't mention much before.

By now, my journey across China with cameras in hand has lasted a decade. I have witnessed the true contemporary moment of China since the country opened the door to the world 40 years ago. The focus of the country has evolved from how to produce more to how to commoditize more. The society has been stressed by uncertainty and sometimes feels split by the past and the future. The value of a product fractured. Objects are derived from one from another. For the designers who are truthful to their creation it has become more difficult to properly judge. It becomes an inevitable and essential question —— What is the truth?

For less than ten years, Dong and Ziying have gained a solid reputation in China and even globally by winning two prestigious awards from the American Society of Landscape Architects. Their work has covered a broad variety in terms of style, size, type, materiality and application of many technologies. Surely, the achievement is impressive, and they are successful. Yet, I could not fully acknowledge that my concern at their return to China was completely wrong. For me, what I know of them now has kept my impression of them intact —— a couple sitting in a beat-up car by the Atlantic Ocean in the winter, quietly painting watercolors. Throughout their projects, I could not finish the puzzle that is supposed to integrate Dong and Ziying into the context of the contemporary China. There are mismatches. There is no doubt that the exceptional circumstances in China are one of the essential keys to their success. Even so, the circumstances can enable even empower but only the designer can embody.

We are living in an era of derivates, but I am sure Dong and Ziying are not looking for the truth for themselves, since their works themselves bear the truth.

Hai Zhang
Feb 14, 2018, New York

前言

现代主义艺术出现的时候（大约以印象派为起点），有一个重要的理念，就是把对绘画作品的观赏转为体验，这也是我理解的传统景观与现代景观之间的差别之一。

现代艺术家们还认为，艺术的工作是展示理念，而不是给人以美感上的愉悦，并认为创造美感是设计师的工作——虽然如此，我理解的设计，身兼传达美与理念双责，同样也是现代景观的职责。简略地说，这两点构成了张唐景观的设计理念。

这本书的目的，是分享一下张唐景观过去十年里，比较满意的十几个作品的详细研究、制作、施工的过程。取名为"静谧与欢悦"，代表设计中不同维度的场景——从安静、温馨，到热闹、欢快。

项目中好看的那面，在公共媒介里都发表过了；它们之所以成为这样的过程，各有曲折惊心。分享的目的，一是满足大家的好奇心，二是同行共勉，看看景观做成这样的效果需要往哪儿努力。总而言之，希望分享的东西对大家有用，对中国的景观行业有用。

还有一些项目，应该说效果平平。我们罗列它们某个精彩的角度，给大家提供一些"意向图片"。其实这些项目也不是不好的项目，只是甲乙双方在深层诉求上的不一致，施工管理不尽如人意，现场违背设计意图等方面多些……虽然我们做每个项目时并没有厚此薄彼，只是实施的过程千差万别。实属事出意外，却又是意料之中。

十年，没有长到可以代表一个时代，又没有短到可以忽略它的存在。此间的两点事情极具时代特征：

第一，快速制造。任何一个项目，只要接手，就要马上出施工图。时代特征，无以为怨。对此我们往往"将设计进行到底"——概念阶段没想好的事，方案再想；方案阶段还没想好，扩初继续……直到施工现场，设计仍有调整的可能。我们有些项目去现场次数会有二十几次。所谓"尽人事，听天命"。

第二，审美粗陋。有人说，一个民族和一个时代的质量往往取决于这个民族和这个时代的审美愿望、审美能力和审美水平……并且，如果因为贫穷，我们在心理上就剔除了美，后果无非就是两条：一、对美的麻木；二、对美的误判。对于审美的讨论虽然牵扯到主观，却因其无所不在、在结果上千差万别而让人寝食难安，因为审美差异的存在不能靠个人努力消除。永远不要让甲方在坏的方案中做选择——即使美学风格不同，仍然有美与不美之分。打磨和提供一个高质量美学的设计，或许是对这个时代最好的应对方法。

Preface

十年，由于行业发展快速，人们对景观新概念的接受度、项目的实施度大大提高。在这里，由衷地感谢为我们提供设计机会的所有业主甲方，由于大家的信任、委托和不断督促，使得这些项目成功落地。

对于张唐的设计师（包括曾经的和现在的），感谢大家在项目中以无穷的耐心面对无尽的修改；在蓬头垢面的加班中为追求更好而做的自我督促；并请在未来的职业生涯中保持设计师的优雅和从容，以包容的心胸和宽宏的心态继续履行一名景观师的使命。

最后，需要特别致谢张唐的设计师姚瑜和王琪。姚瑜为《参与性景观》和本书的出版花费了大量时间，联系出版、研究版面、反复订正。她们对排版不厌其烦的考究，以及个人精湛的审美功力，为此书做出了极大贡献。也为关心这次出版，为这两本书的诞生出谋划策的朋友们表示由衷感谢。

我们的摄影师张海是多年老友。每次拍摄都要从美国飞到中国，一年数次，可谓辛苦。也因此他目睹了张唐十年来的项目经历。对于我们的设计，他从艺术家的角度有自己的理解。我们一直引为知己，并非常荣幸地请他为本书写了序。至此，言终。千钟粟自在书中。

唐子颖
2018 年初春，上海

When Modernism in the arts began approximately with the emergence of Impressionism, a fundamental change had taken place —— art was now meant to be experienced, not just looked at.

This distinction is critical to understand the difference between traditional and modern landscape architecture.

Modern and contemporary artists have dispensed with the notion that the arts should provide purely aesthetic joy. Now art has become a medium of expression of the artists' concepts and theories. They considered an artist's job not to provide aesthetic pleasure that was the job of the designers. In my opinion, the designers, in fact, should bear a dual-responsibility —— to provide aesthetic joy while also presenting the artists' ideas.

This dual-responsibility has become the design philosophy for Z+T Studio.

The purpose of this book is to share. We wanted to share our research, the process of design, fabrication and construction through the example of dozens of representative projects. As we wanted these projects to illustrate various dimensions of our design —— from quiet and sweet to dynamic and lively, the book is entitled —— The Tranquility and Joy.

There are many aspects to every project. In the past, many beautiful images of the projects have been shared in magazines, journals and on social media. Here, we want to take the opportunity to share more stories from the process to realize these projects. We hope it might 1) offer more information to those interested in our design; 2) most importantly, be a force to encourage and inspire introspection among peers. It is by examining the past that we hope to better understand the future. All in all, we hope the contents of this book could assist readers as a humble contribution to the study and understanding of landscape architecture in China.

Some projects in the book may appear ordinary upon their completion. Yet, in the book, we would like to share a few noteworthy angles of these projects to illustrate "ideals". These projects all sprung from interesting ideas. The end result may have been different from the original intention for several reasons : the clients and the designer may have aspired differently or else some contractors may have been uncooperative and imposing. We treat every project in our studio equally, but the results after the construction have varied greatly. The final appearances of many projects were largely unexpected.

A single decade is not long enough to represent an entire era, but it is significant enough not to be ignored. For this decade, two things perfectly illustrate our time.

The first is the obsession for a fast pace. For nearly every project in our studio, the construction drawings were demanded as soon as the design process had commenced. Despite the character of our era to always be moving at a rapid pace, we never give up the commitment to the essentials of the design. The truth is that we have carried many unsettled aspects and ideas of the design into later design phases —— some conceptual design in the schematic design phase, some schematic design phase work in the design development phase, and so on. The design itself could well continue onto the construction site. It is not uncommon that we might hold dozens of on-site design sessions and workshops during the construction. We believe that we should always give our best effort and hope for the best result.

The second is lack of the aesthetic taste and judgment. There is a saying that the quality of a nation and time can be measured by the nation's aspirations and aesthetic judgment. Economic goals would never be a reason to abandon our aspiration for beauty even though it persists as an impulse instead driven by our insensitivity and misunderstanding of true beauty. Although aesthetics is largely subjective, it relates to almost everything, especially design. To a designer, it could be extremely troublesome. We understand the impossibility of completely removing differences in aesthetic tastes, yet we will never present a design that might appear to compromise our standard. As we acknowledge differences in aesthetic appreciation, we believe it just as obvious whether the project is beautiful or not. To commit to the high standard probably is the best way to contend with this imperfect era.

For this decade, the level of acceptance to new concepts and the quality of construction in China have improved significantly. Here, given a chance, we would like to express our sincere appreciation to the clients who have trusted us, granted us opportunities, and encouraged us. Without them, realizing these designs would have been impossible.

We also want to thank the designers, current and former, in our studio, for their infinite patience to the endless revisions, being self-motivated for the better design. We hope they can maintain the grace and composure while being forgiving and open-minded in their continuing career as landscape architects.

A special acknowledgement goes to our designers Ms. Yu Yao and Ms. Qi Wang. Ms. Yao has spent enormous time and effort on editing and logistics for these two books. Ms. Yao and Ms. Wang revised the design of these books numerous times. We appreciate their great sensitivity and talent. We thank all friends who have contributed ideas and expressed interests in the books over time.

Finally, we want to express our gratitude to our dear friend, Mr. Hai Zhang. He is the photographer who visually captured most of our projects. He has come to China from New York where he resides for much of every year. As an artist witnessing and interpreting our projects, we consider his photographs not only the documentation but also artistic, providing a truthful interpretation of our projects and the context. We greatly appreciate his support and friendship and feel privileged having him write the foreword for this book. Let me stop right here, and let the book speak for itself.

Ziying Tang
2018, Spring, Shanghai
Translated by Hai Zhang

目 录 | CONTENTS

12　序：寻找衍生品时代的真实
　　Foreword : Looking for the Truth in an Era of Derivatives

16　前言
　　Preface

22　樾园
　　Yueyuan Courtyard

48　永泰会所
　　Yongtai Club

54　玖著里
　　Jiu Zhu Li

74　富力十号
　　Royal Territory

80　九里云松
　　Pins De La Brume

102　玉湖会所
　　Yuhu Club

110　建研中心
　　Vanke Research Center

124　京华园
　　Jinghua Garden

132　山水间
　　Hillside Eco-Park

162　良渚文化村
　　Liangzhu Village

174　嘉都公园
　　Jiadu Park

194　鲸奇谷
　　Marvel Valley

214　CMP 广场
　　CMP Plaza

220　云朵乐园
　　Cloud Paradise

252　公园里
　　The Park

278　理想城
　　Dream City

284　东方传奇
　　Oriental Legend

290　优盛广场
　　U-Center Plaza

310　附录
　　Appendix

樾园
Yueyuan Courtyard

苏州樾园尝试运用当代材料与工艺对苏州古典园林进行现代演绎。景观设计没有停留在对古典园林的浅层视觉呈现，而是通过对自然元素的抽象表达，传达更多言外之物。最为核心的创意是与景观密切结合的雕塑水景，以极高的施工完成度巧妙地与环境相融合，创造了一个独具特色的现代文化景观。

樾园的设计源自对苏州园林经典美学的再理解。水景作为中国古典园林的重要元素，在樾园中作为场地的一条视觉主线，连接了两块核心区域——"溪院"与"湖院"。访客循着蜿蜒浅溪进入"溪院"，主题水景通过对整石花岗岩纹理化的精巧加工，象征了山川河流历经千百年的奔流不息，将平坦的陆地侵蚀成河谷的过程。当浅溪蓄满水，形成一个曲折镜面倒影池，既暗含了长江的形态寓意，也借景反映了瞬息万变的天光云影。这条有如从地面雕刻出来的折形倒影池，结合建筑中庭投射的光影与潺潺溪流声，营造出了静谧、冥思的空间氛围。浅溪的尽端，溪流最后汇入作为"湖院"主景的另一个倒影池。动与静的体验经过简约、巧妙的处理，在这个不到 1 000 平方米的庭院里达成了理想的融合。

在施工过程中，最大的挑战是如何将这个复杂的曲线水景雕塑与景观无缝衔接。在景观设计师与艺术工作室成员的同心协力下，形态复杂的水景石雕被切分成数个片段单独加工和制作，然后在现场进行组装，最终使水景雕塑与周围环境和谐统一。不同于传统的苏州古典园林，樾园通过极简的植物配置与石材的运用，体现了当代景观设计对材料处理的精炼手法。樾园景观的独特性来自对水景、石材、植物等景观元素的挑选和恰如其分的运用。置身庭院中，流转的光阴、更替的晴雨与景观空间的动态呼应，结合现代感的细节与花岗岩石刻上的流畅曲线，为人带来了一种截然不同的空间体验。

The Yueyuan Courtyard, located in Suzhou, showcases a landscape expression of modern architectural motifs and understanding of the rich culture and history in classical garden making. Though inspired by the local Suzhou classical garden, the landscape design does not focus on simple representations. The design abstracts the elements from cultural traditions and nature. The most innovative design is the sculptural water feature which is integrated into the landscape artfully and serves as a tribute to the region's distinctive natural characteristics along with its tradition of fine craftsmanship.

The water feature weaves the elements in the courtyard together and connects the two main parts of the Yueyuan Courtyard, the Creek Garden and the Lake Garden. The visitor's experience begins at the Creek Garden, and moves along to the Lake Garden from the creek that inspired the stone sculpture. The water feature, made from granite monolith, artfully simulates the natural hydrological processes — how the runoff water moved over the surface of the earth for millions of years, transforming the land into river valleys. When filled with water, the creek becomes a curved reflecting pool. The meandering pool reflects the environment and frames artful images on the water. The effect created by this sculptural water feature is metaphor of scenery-borrowing along with an ever-changing sky, a poetic description of the Yangtze River.

As the light pours into the central garden and time passes, the people could experience the moving shadow and the sound of the water flowing through the creek. It generates a tranquil, meditative atmosphere. At the end of the creek, the water flows into another pool in the lake garden. Z+T Studio created a simple yet complex experience in such a small natural space. Contrasting to the Suzhou classical gardens, the Yueyuan Courtyard's palate of plants and stones is minimal. It expresses a distillation of the materiality and the contemporary artistic mind.

The project is technically challenging — how to embed a sculptural water feature with a complex geometry into the courtyard. With a collaborative effort from the landscape architects and the craftsmen from the art workshop, the sculpture was completed in segments. They were fabricated individually and then installed carefully on site. It relied on a series highly precise movements throughout the entire process.

△ 苏州古典园林 | Suzhou Classical Garden

▷ 项目区位 | Project Location

▽ 场地现状 | Existing Condition

樾园 Yueyuan Courtyard / 25

草图研究 | Sketches

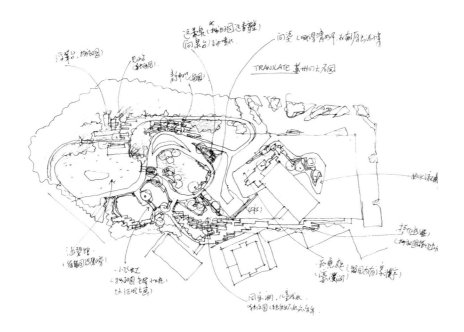

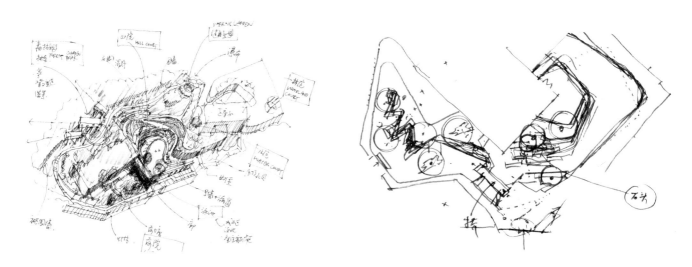

26 / 静谧与欢悦 Tranquility & Joy

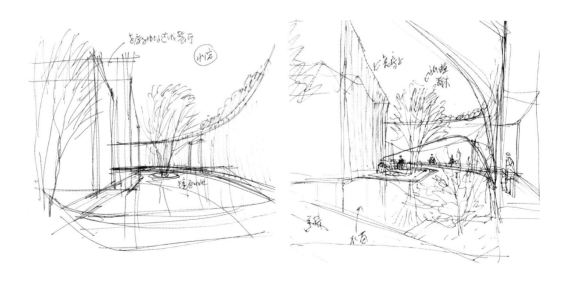

樾园 Yueyuan Courtyard

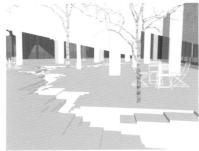
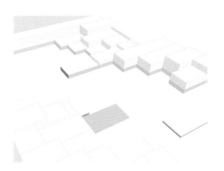

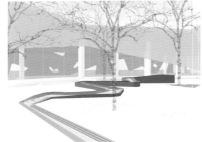

28 / 静谧与欢悦 Tranquility & Joy

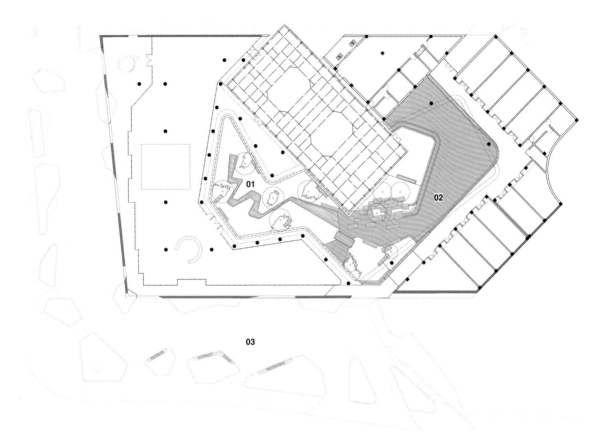

01 溪院 Creek Garden

02 湖院 Lake Garden

03 外街 Promenade

△ 总平面 | Master Plan
◁ 研究过程 | Modeling Research

施工过程 | Construction Process

溪院 | Creek Garden

樾园 Yueyuan Courtyard / 37

溪院细节 | Creek Garden Details

樾园 Yueyuan Courtyard / 39

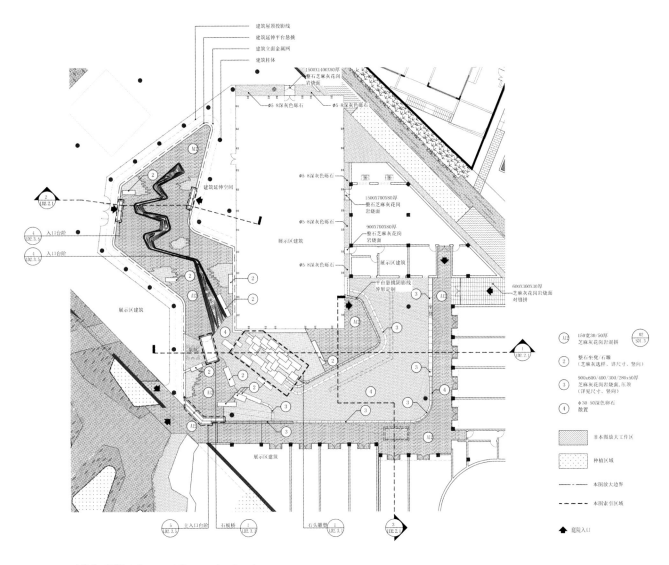

庭院施工图纸 | Courtyard Construction Drawings

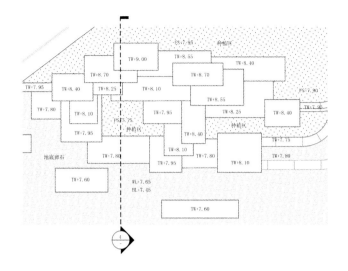
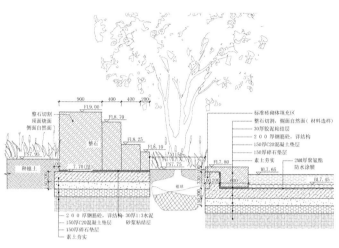
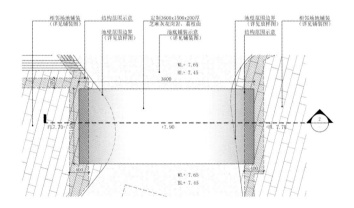
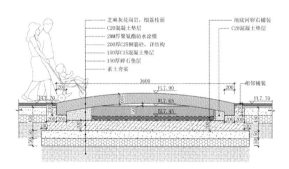

樾园 Yueyuan Courtyard / 41

湖院 | Lake Garden

樾园 Yueyuan Courtyard

永泰会所
Yongtai Club

该商务会所位于上海西郊宾馆一角，总用地面积大约1.5公顷，其中包括3栋高端奢侈品展示会所，2栋商务办公会所和1栋休闲会所。场地内原有7栋荒废多年的别墅，建筑师将其改造为带有古典韵味的建筑形式。景观改造没有简单地跟随建筑风格，而采用相对去风格化的方式，将关注点放在对整个园区的空间格局、氛围、使用和景观感受上。在有限的空间里既考虑公共共享空间的组织和系统性，又强调每一个别墅会所私有庭院的独特性和私密性。现代简约的硬质景观配以简约的植物景观，使用耐久的景观材料，考虑永久性的设计手法，用细节体现了高端社区现代景观的基调，营造浪漫与严谨相交融的氛围。

The Yongtai Club is a part the Shanghai Xijiao Guest House. The club occupies a site of 1.5-hectares. Prior to the landscape design, the client commissioned an architect to renovate 7 abandoned mansions for the projects. The buildings were designed and renovated in a highly decorated colonial style. The landscape design took a contrasting approach and focused on creating the space, atmosphere and functionality. In the limited space, the design re-organized the shared space throughout the site, at the same time preserving the privacy of each mansion. Aesthetically, simplicity and elegance ruled the selections of the material and plants. Refined craftsmanship was central to the project.

△ 涌泉 | Bubbling Spring
◁ 水景墙 | Water Wall

玖著里
Jiu Zhu Li

玖著里位于宁波，是住宅建筑和街道之间的一块三角形地块，面积约 5 000 平方米。会所建筑包括书院、小型餐厅、健身等功能，需要设置次入口，未来有对外直接开放的可能性。景观设计师和建筑师密切合作共同确定了社区入口、会所建筑的位置以及整体空间格局。

景观设计受宁波当地著名文化景观天一阁的启发，探索在现代城市中营造符合当代生活、习惯和审美，同时又有古典韵味和传承的景观。在设计过程中，我们尽量避免直接采用古典园林中符号化的元素，而更多地从空间序列和自然感受上着手去把握古典园林的精神和内涵。希望通过静谧的园林氛围，引人内省。对于玖著里来说，园林中的自然，既不是古典园林中的微缩山水，也不是现代生态景观中的乡野田园，而是自然中更本质的元素：时间、光、影、水、木、土、石。

玖著里的主要景观包括入口广场、南侧连廊、天一水苑、静水长池及光影长廊。设计着重关注各个空间的尺度、边界以及空间之间的转折衔接关系。项目首次尝试 U 形玻璃作为室内外隔断的材料，其透明度、颜色等的选择与应用在设计中起了至关重要的作用。

Jiu Zhu Li is an exclusive club that houses a library, reading rooms, fitness center and cafeteria in Ningbo, Zhejiang Province. It occupies a 5,000 ㎡ triangular space between residential buildings and the municipal street.

The landscape design of Jiu Zhu Li was inspired by Tianyi Ge (One Sky Pavilion) — a famous library with gardens built during the 16th century in Ningbo. Z+T Studio was exploring a design that could fit in the context of the contemporary urban life, yet continue the lineage of the strong classic culture and landscape designs of the region. The essence of the regional landscape was learned and transformed rather than being replicated as the symbols. The meditative atmosphere and spirit of the classics were embodied through spaces carved out in Jiu Zhu Li. The landscape creates an introspective space in which nature is experienced and appreciated. In Jiu Zhu Li, nature is treated beyond an object. It is neither replicated as a miniature in the classic garden nor a wild artificial field. By integrating the essential elements — time, light, shadow, water, wood, earth and rock, Jiu Zhu Li has become a space that unites these elements and the man-made environment into one humble yet inspiring experience.

Jiu Zhu Li consists of the entry courtyard, the south galleria, the Tianyi water court, a series of linear reflecting pools and the light-and-shadow galleries. The design pays close attention to the proportion of the spaces, its transitions and articulations. The selection of the materials was carefully studied and included an array of natural stones, metal, translucent U-shape glass and frosted glass panels.

◁ 场地现状 | Existing Condition
▷ 草图平面 | Sketch
▽ 总平面 | Master Plan

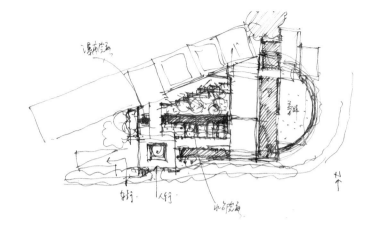

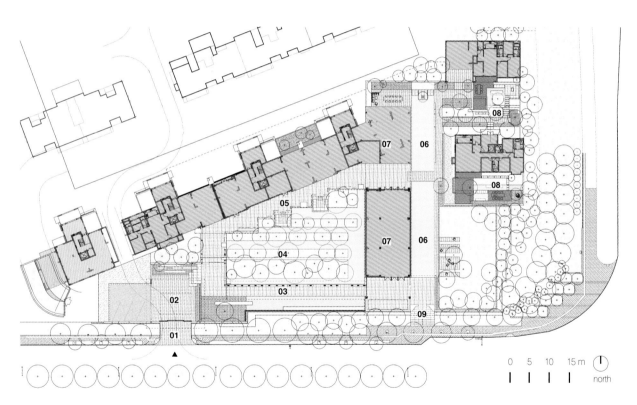

01 主入口 The Main Entrance
02 入口广场 Entrance Plaza
03 南侧连廊 The South Corridor
04 天一水苑 Tianyi Water Garden
05 光影连廊 Light and Shadow Corridor
06 静水长池 Quiet Water Pool
07 会所 Clubhouse
08 样板院 Model House
09 次入口 Secondary Entrance

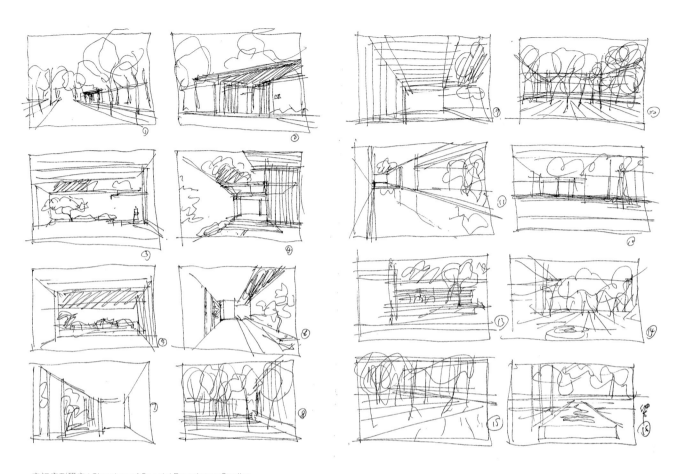

空间序列研究 | Sketches of Spacial Experience Studies

草图研究 | Sketches

玖著里 Jiu Zhu Li / 59

60 / 静谧与欢悦 Tranquility & Joy

施工过程 | Construction Process

玖著里 Jiu Zhu Li / 61

入口格栅 | Entrance Steel Screen

入口庭院 | Entrance Courtyard

U 形玻璃屏风 | U-Shape Glass Screen

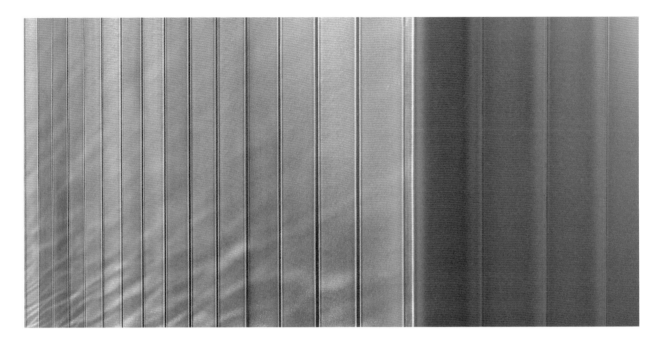

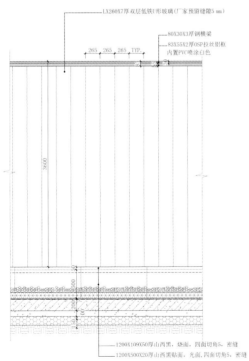

△ U 形玻璃屏风细节 | Details of the U-Shape Glass Screen
◁ U 形玻璃屏风施工图 | U-Shape Glass Screen Construction Drawings
▷ 南侧连廊 The South Corridor

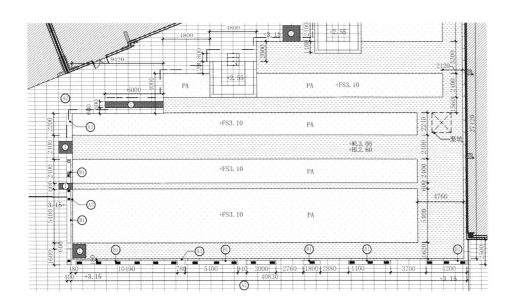

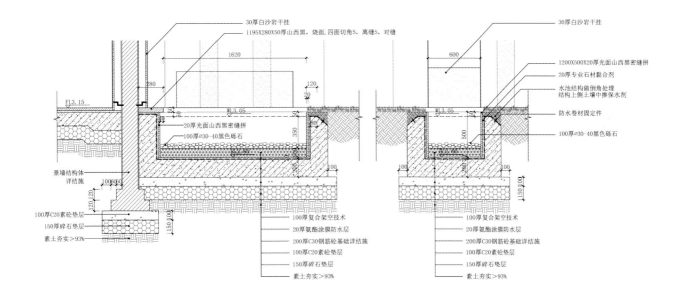

△ 天一水院施工图 | Water Garden Construction Drawing
◁ 天一水院 | Water Garden

光影连廊 | Light and Shadow Corridor

富力十号
Royal Territory

富力十号位于杭州市以西，临近西溪湿地，离杭州市中心直线距离约13千米。其中A块场地周边多为建设工地。为营造静谧、舒适的空间环境，景观设计师设计了高低错落的墙体，进行空间组织和围合的同时兼顾沿街立面的整体形象。景观空间分为四个部分，空间序列的设计构成了步移景异的效果：入口以一道7米宽的水幕将行人引入庭院，通过舒适的林下幽径、泰山石石台、开阔的镜水面，穿过建筑门廊后停留在一处林荫庭院，然后进入社区会所。

B地块景观用紧凑的空间结构和简约、诗意的设计元素，在狭小的空间里体现公共空间的仪式感，同时兼顾院落的私密性。景观中轴的《富春江流域》水景雕塑是安装在地面上的一片倒影池。这是张唐景观第一个尝试在石材上做水纹雕刻的项目。我们提取钱塘江及周边山脉的地形纹理，转化成具有艺术感的铺装肌理，并借由水位的高低变化，形成水盈则为镜面倒影池，水亏则为山峦沟壑或潺潺溪涧的浮雕水景。

The Royal Territory is located in Hangzhou close to the Xixi National Wetland Park. The project includes sites A and B.

Site A is at the entrance of the project and directly faces the busy construction surroundings. Z+T Studio utilizes a series of walls of various heights to create several in-between spaces to separate the inside from the outside chaos. The walls also create a unique identity for the project to stand out from the surrounding. The principle of the design is to create various scenarios with different social experiences. Through the waterfall screen at the entry, the alley of tree shade, the Taishan stone and a reflecting pool, the courtyard in the tree shade is finally revealed behind a steel screen.

Site B is a courtyard separated from the outside environment, yet very linear. In its limited space, Z+T Studio introduces a relief-like stone sculpture through the central axis of the courtyard. The pattern of the sculpture was inspired by the local geography and abstracted from the topography of the Qiantang River and surrounding mountains. With its subtle slope, the water flows from one end to the other. The slow movement of the water through the sculpture creates a poetic miniature scene of the river flowing through gorges. The reflection creates the illusion of a larger space. The pattern of the sculpture also resembles a piece of relief art.

△ 中心庭院 | Central Courtyard

◁ 水景雕塑细节 | Relief Fountain

△ 入口庭院 | Entrance Courtyard

▷ 中心水院 | Central Courtyard

九里云松
Pins De La Brume

　　九里云松改造项目位于杭州西湖风景名胜区，毗邻西湖，背靠千年古刹灵隐寺。原有酒店建筑和庭院已破败不堪。改造工程由于受景区规范限制，需要保护现有大树和主体建筑结构，并在材料、色彩和设计上体现传统风貌。

　　酒店外部空间分为入口庭院、中央核心庭院、餐饮区庭院、泳池区庭院以及屋顶庭院五个部分。景观设计在处理功能流线、保护现状大树的基础上，着重考虑如何通过现代极简的手法表达中式庭院的起承转折，营造安静优雅的休闲气氛。通过对光线、光影、倒影、流水等细节的研究和设计，达到人与自然的亲近交流。中央核心庭院充分考虑了现状大树的位置、标高以及和建筑室内功能的关系，通过方形倒影池、汀步、叠水景墙、草坪灯景观元素确定基本空间形态。四片屏风式景墙将主庭院中的一部分划分为餐厅包厢的外部庭院，使得内外两个庭院之间隔而不断。

　　该项目是对户外屏风墙在景观中的首次尝试。景观材料的选择尽量与建筑立面改造以及周围环境相协调，其划分方式与墙体的高矮、长短比例相得益彰，在业内传为经典。

Pins De la Brume is a luxury boutique hotel built on the site of a mostly abandoned guest house and gardens. It is located in the famous West Lake National Scenic Area in Hangzhou, Zhejiang Province and adjacent to West Lake, the Xixi National Wetland Park and the 1,000-year old Lingyin Buddhist Monastery. For such a prestigious and historic location, it was decided that the design of the new hotel must follow rather strict regulations so as to preserve many of the historical trees, the structure of the main building on the site and the integration of the color, style and material of the surrounding architecture in the area.

In Pins De la Brume, the core of the outdoor space consists of five courtyards — the entry courtyard, the central courtyard, the dinning courtyard, the swimming pool courtyard and the rooftop courtyard. As all programs and spaces are organized by these five courtyards, the transition between the yards is essential to the overall landscape design. Z+T Studio has created a design that both reflects the influence of the traditional Chinese courtyard yet remains essentially modern. The design of the central courtyard creates a synergy between the site's history and the requirements of the new hotel. The historical trees are considered as the pivot of the organization as the reminder of the past and the lead to the present. Through a series of square reflecting pools, traditional stone pathways through the water, a falling water wall and lawn lighting, the interior and exterior features are all unified. The four screen walls offer privacy for the VIP dining space without containing the courtyard. On the contrary, the screen walls expand the optical boundary of the space.

The selection of the materials harmonize with the surrounding architecture and environment. The color palette of the material is minimal — grey granite for the pavement, dark grey granite for the falling water wall and a light color granite for the screen walls.

◁ 场地现状 | Existing Condition
▽ 草图平面 | Sketch

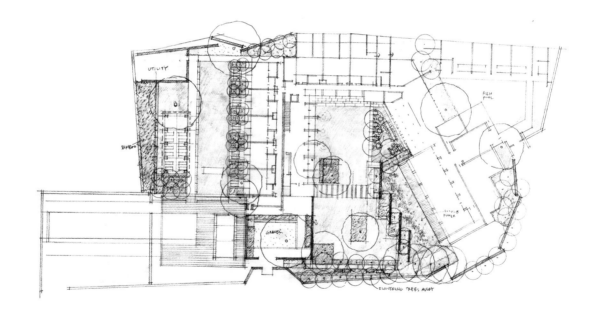

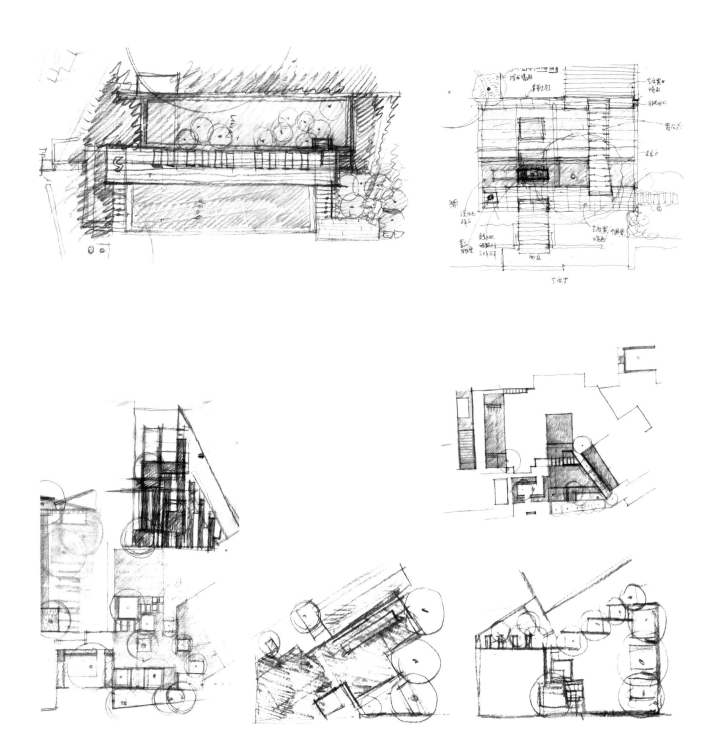

84 / 静谧与欢悦 Tranquility & Joy

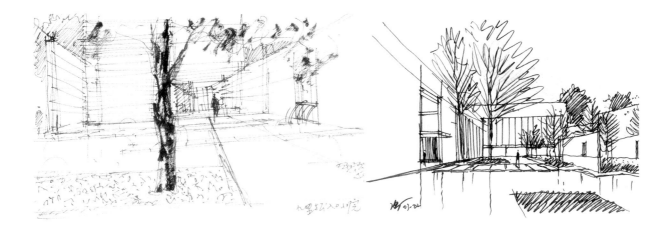

草图研究 | Sketches

九里云松 Pins De La Brume / 85

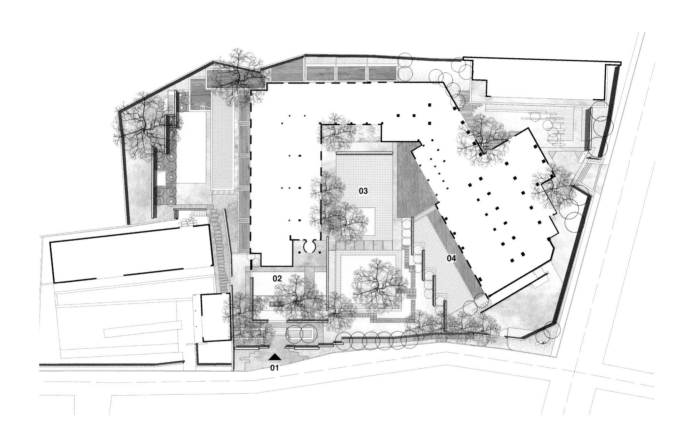

01 主入口 Main Entrance
02 入口庭院 Entry Courtyard
03 主庭院 Main Courtyard
04 餐饮区庭院 Dining Area Courtyard

◁ 总平面 | Master Plan
▽ 轴测图 | Axonometric Drawing

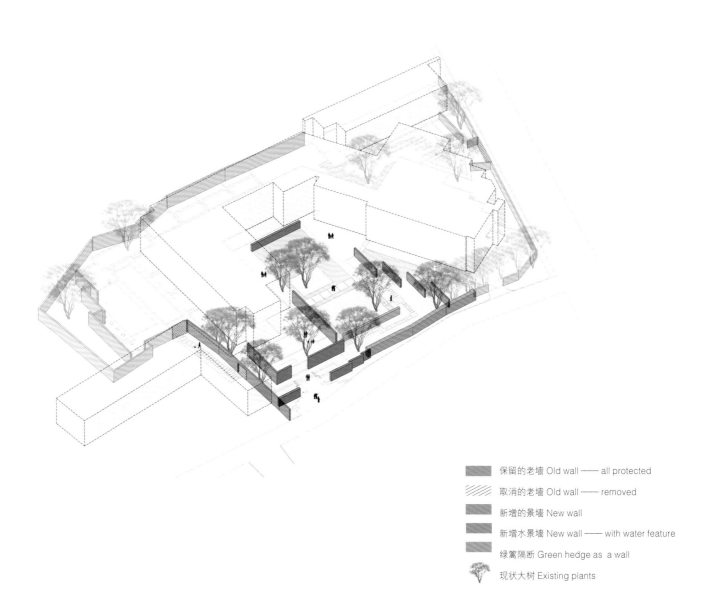

保留的老墙 Old wall —— all protected
取消的老墙 Old wall —— removed
新增的景墙 New wall
新增水景墙 New wall —— with water feature
绿篱隔断 Green hedge as a wall
现状大树 Existing plants

九里云松 Pins De La Brume / 87

施工过程 | Construction Process

88 / 静谧与欢悦 Tranquility & Joy

九里云松 Pins De La Brume / 89

入口庭院 | Entry Courtyard

△ 砂岩屏风墙 | Sandstone Screen Wall
▷ 景墙施工图 | Screen Wall Construction Drawings

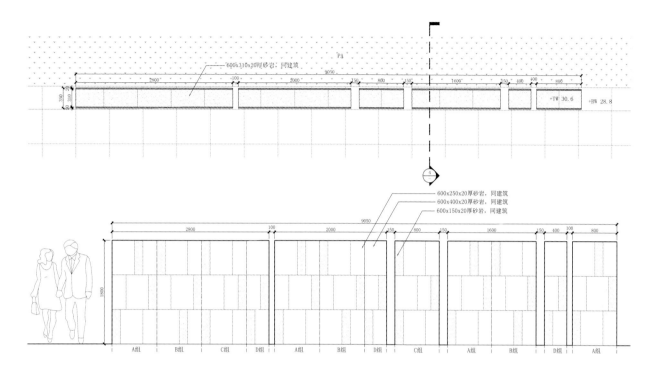
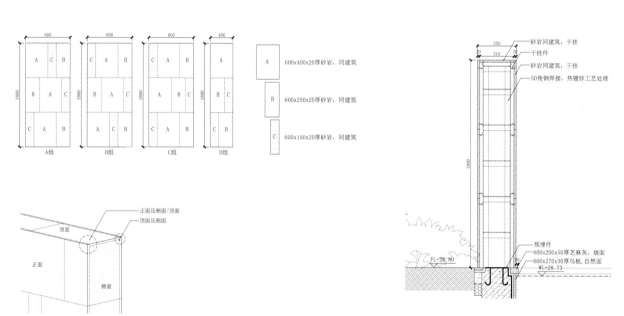

九里云松 Pins De La Brume / 95

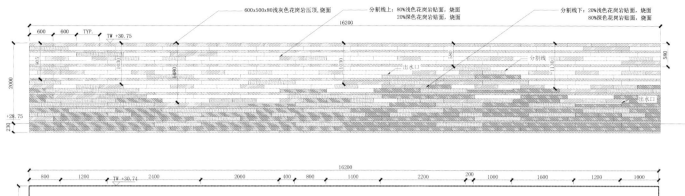

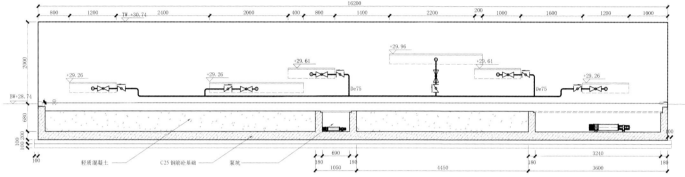

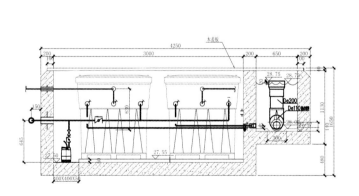

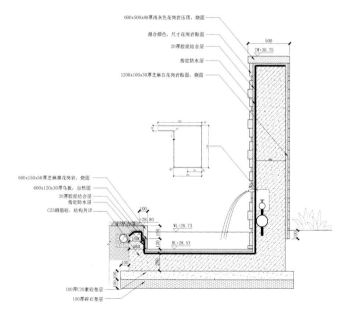

△ 水景墙施工图 | Water Wall Construction Drawings
◁ 水景景墙 | Water Wall

九里云松 Pins De La Brume

水院细节 | Water Courtyard Details

玉湖会所
Yuhu Club

玉湖会所位于昆山市郊外临湖半岛上，场地相对狭小、独立，面湖一侧环境优美，其他朝向均为建设用地，比较杂乱。景观设计充分考虑了从入口停车场到会所的空间感受，通过林荫道、石材景墙、水景和亚克力半透明屏风对前院空间进行梳理，将杂乱的外围环境与会所做适当隔离，让人逐渐地从嘈杂的场所进入安静的环境。会所建筑临湖空间布置错落的户外活动平台与相应的室内空间连接，形成多样的临水活动场地。平台和湖面之间种植茅草、狼尾草、千屈菜等当地水生、湿生植物，提供活动平台和6米深水面之间的安全区域，同时形成有特色的滨水景观。

该项目是张唐景观首次尝试使用亚克力作为户外屏风的材料。作为半透明的隔断，该元素布置在建筑入口主空间和后勤服务空间之间。设计灵感来源于中国传统室内灵活划分空间的屏风。半透明的材料可以捕捉到阳光的变化、植物的投影，并使两个空间隔而不断，形成意味深长的联系。

The Yuhu Club is located on a lakeside peninsular outside of Kunshan, Jiangsu Province. As the three sides of the project were surrounded by the construction site, the lake offers great views on one side for the project. The landscape design creates a sequence to organize the spaces from the parking lot to the main building of the club. Composed of a tree-shaded alley, stone walls, a water feature and acrylic translucent screen walls, the spatial sequence is the prelude of the project. It gradually reveals the design and screens from a surrounding environment that seems more chaotic. A series of platforms at various elevations were placed on the outdoor space by the lake, bringing the scenic lake view into the main building and merging the outdoor and indoor spaces together. Along the shore of the lake, couch grass, Chinese fountaingrass, lythraceae and other local aquatic plants were planted, providing a safety buffer to people on the platforms while creating a unique lakeside scene.

It was the first time that Z+T Studio utilized acrylic to create outdoor features. The idea was inspired by the screen wall in the traditional Chinese architecture that both defines the spaces while maintaining spatial continuity. For Yuhu Club project, the translucent acrylic screen walls are used as a partition between the space at the main entry and the service area. The translucency of the acrylic material captures and reflects the subtle changes of sunlight, shades and the movement of the tree branches. The effect of shadows cast onto the acrylic screen resemble an abstract Chinese painting.

植物细节 | Plant Details

建研中心
Vanke Research Center

建研中心是万科集团研究住宅产业化的基地园区，包括研究室、厂房、实验室、生产车间以及部分生活配套用房，主要进行建筑材料技术、低能耗、低碳环保相关方面的研究。张唐景观负责园区中几个分散的场地景观设计工作。景观设计符合研究中心的整体理念，探讨在景观领域中的各种环境友好型策略，包括雨水管理再利用、生态材料研发、预制混凝土模块在地产项目中的应用、透水材料，以及低维护策略应用等。此外，通过这些尝试探索如何将景观的生态与艺术结合起来，使生态景观成为可供欣赏、教育和参与的场所，便于被更多的人接受，融入日常生活。

由于位于研究中心，该项目是动态的，可供观察、记录、调整和修改。我们希望可以借此摸索出一套适宜于中国当前的技术、经济状况的低能耗生态景观设计方案。项目于2010年正式启动，2012年大致完工。根据用地条件主要包括三个方面的内容：预制混凝土模块的研发与应用；景观生态水循环处理系统的展示；景观生态材料与手法的试验与应用。项目通过预约可对外开放供游人参观，便于传播生态理念。

The Vanke Research Center is a campus dedicated to the advanced architectural and construction research of residential buildings for the Vanke Group, one of the leading real estate developers in China. The campus houses the research labs, workshops, testing labs, and service facilities. The research conducted at the center focuses on advanced and sustainable building materials, energy-efficient design, low carbon design and overall sustainability. Z+T Studios was commissioned to design the landscape of several sites on the campus. Rather than a regular landscape design project, it has become part of the research as well as a lab. As it integrated the research design, the project has explored the strategies of rain water control and reuse, eco-friendly materials, the implementation of prefab concrete and low-cost maintenance. The project additionally explored possibilities for making ecologically-responsible landscape educational. It encourages the participation of the users to bring awareness of the environment.

As the project has become a part of the research in the Vanke Research Center, it is ever-evolving. The project could be scientifically observed, recorded and analyzed. The design has offered the flexibility to adjust and modify the landscapes accordingly. The goal is to provide a living lab to research low-cost and low-maintenance landscape design strategies and technologies suitable to the specific ecological and economic environment in China. The process of the design and implementation of the project spanned two years from 2010 to 2012. The research center is open to the public by appointment.

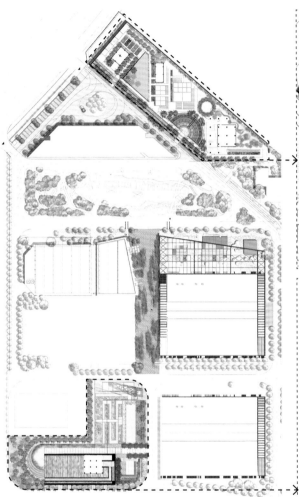
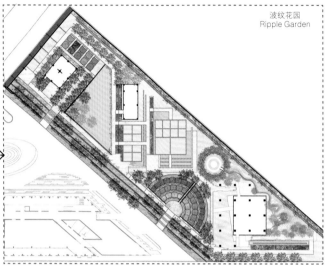
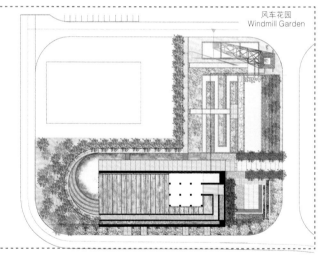

总平面 | Master Plan

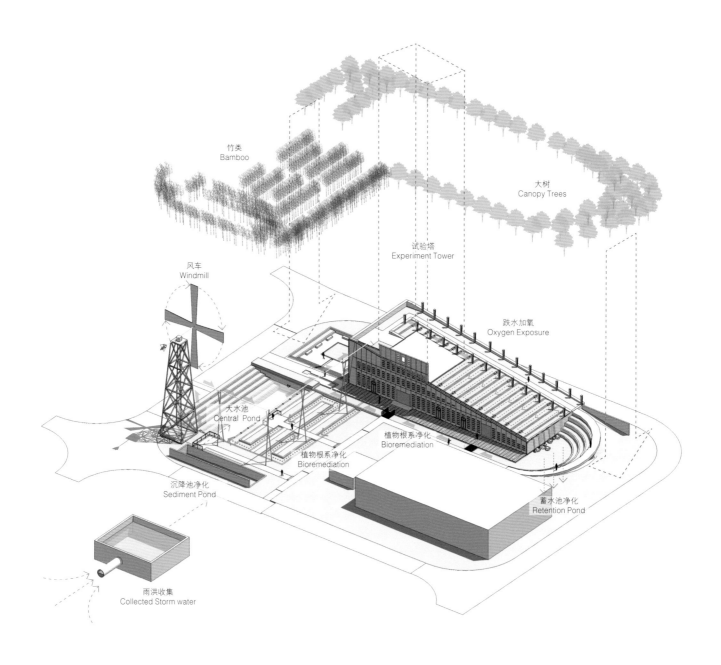

△ 净水系统图解 | Water Treatment System Diagram
▷ 试验塔 | Experiment Tower

△ 半环波浪区 | Semicircle Ripple Garden
▷ 植物净化池 | Bioremediation Tank

嵌草预制混凝土细节 | Precast Concrete Details

京华园
Jinghua Garden

京华园是2009年北京花卉博览会北京市的展示园区，面积约5 000平方米。展示主题为"京华双娇，古韵新妆"，需要突出展现北京市的两种市花——月季和菊花。根据场地特征和要展示的花卉内容，张唐景观与北京花卉园艺研究中心合作，将园区分为入口雕塑广场、菊花展示区、藤本月季展示区、月季台地区、岩石园和建筑室内展示区等。各个区域的设计构思都着重于对北京市花菊花和月季的充分表达，并始终渗透到景观的形态与材料之中，形成鲜明的北京特色，集中体现了月季和菊花在京城广泛种植的情况。园中异形景墙取意于月季花瓣，使月季展示区的花坛极具特色；抽象的菊花形的钢桥，以曲线的柔美，营造似穿行于花瓣中的意境；菊花形流水将弧线的自然美与水流的灵动巧妙地结合在一起，成为游人可以参与、嬉戏的场所。

The Jinghua Garden is a 5,000 m² showcase garden which represented the city of Beijing during the 2009 Beijing Horticultural Expo. The garden features the two city flowers of Beijing — the Chinese rose and chrysanthemums. Z+T Studio designed the Jinghua Garden in collaboration with the Beijing Horticulture Research Center. Constrained by the existing condition of the site and the requirements of the exhibition program, the garden was divided into multiple areas — the sculpture plaza at the entry, the chrysanthemum garden, the climbing Chinese rose garden, the raised platforms for the Chinese rose, the rock garden and the indoor exhibition pavilion. To showcase the varieties of Chinese rose and chrysanthemums is one of the main missions of the design. The design of the garden demonstrated the broad applications and deep involvement of these two types of flowers in people's daily life in Beijing. Z+T Studio also designed a few functional features as the geometric abstraction of the organic forms of the flowers. Both the double-curved steel bridge and the creek were inspired by chrysanthemums.

△ 板岩花池 | Flowerbed

◁ 菊花桥 | Chrysanthemum Bridge

山水间
Hillside Eco-Park

　　山水间位于长沙，是一个典型的中国高密度社区里的公共绿地。它被四周超高层住宅包围，为新入住的几千名住户提供户外活动空间。公园面积为1.4公顷，需要满足各类人群的不同使用需求。场地位于整个社区的低洼点，内部有大片可保留的山林和一个鱼塘。景观设计尊重现有自然资源，尽量保护现有植被和水文，为满足居民日常生活需求设计系列活动空间和场所，将自然和活动交融一体。雨水管理系统和场地相辅相成，并通过系列环境教育标识和设施，使人的活动和雨水生态系统产生互动。孩子们在玩耍的同时也可以了解到与雨洪生态相关的知识。

　　在现有山林和池塘之间，利用原有地形设置儿童活动场地。滑梯在原有坡地上改造而成；木平台剧场受场地内原有林地间系列小台地启发，形成趣味林下休息空间；符合场地气质的"大昆虫"主题的引入，让活动场地更加独特，增强了可识别性和可记忆性，让前来玩耍的小朋友留下独特而美好的记忆。

　　山水间是张唐景观里程碑式的作品。在项目里，我们首次尝试了完整的雨洪系统在社区公园的应用，并将生态教育贯穿其中；第一次设计研究波浪形木平台在户外的应用；第一次使用钢筋网编制户外家具、互动雕塑。最后，项目以极高的完成度呈现，对行业发展起到了深远的影响。

1.4 公顷社区公园
1.4 hectares of community parks

△ 场地现状 | Existing Condition
◁ 场地分析 | Site Analysis
▷ 草图平面 | Sketch

134 / 静谧与欢悦 Tranquility & Joy

The site condition of the Hillside Eco-Park is typical for the high-density residential communities in urban China. The site is surrounded by high-rise residential buildings and the park serves hundreds of new residents in the community. Although the park only occupies 1.4-hectares, the park must meet a wide range of requirement as it serves different groups of users. Topographically, the site is at the low point of the community. A large grove of trees stands on a hill overlooking a fish pond. The core principle of the design is to preserve the existing ecosystem including the body of water and vegetation as much as possible. The program is designed to provide diversified activities for the residents as well as to integrate with nature. While a water management system was incorporated into the design of the park, the park has also become an educational center for sustainable uses of water resources. Z+T Studio designed a signage system to introduce the specific knowledge related to the water and environment. Interactive playground equipment was designed and fabricated by Z+T Studio for the park. Through participatory play activities, children can learn with memorable experience.

Following the existing topography, the space at the foot of the hill and in front of the fish pond hosts the park's primary playground area. The sloping side of the hill offers perfect opportunities for slides. Taking advantage of the elevation changes, a series of stages and resting platforms in the shade of the tree grove was designed. On the lawn that slopes up from the pond to the playground, Z+T Studio designed, fabricated and installed several gigantic interactive sculptures of ants. The magnified size and specific design of the sculptures offer children an unusual experience while further connecting their play to nature. They have since become one of the visual identities of the park.

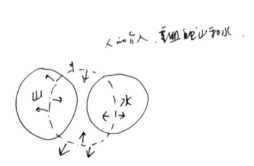
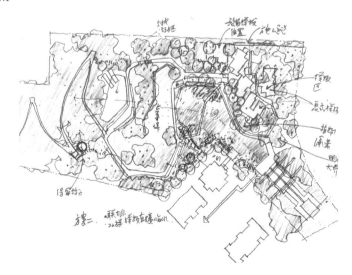

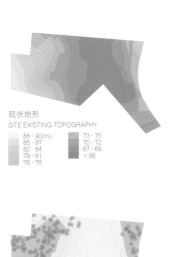

现状地形
SITE EXISTING TOPOGRAPHY

88 - 90(m)　　73 - 75
85 - 87　　　 70 - 72
82 - 84　　　 67 - 69
79 - 81　　　 < 66
76 - 78

场地中央具备收集四周雨洪的潜力
The central area is a potential place to gather storm water from surrounding hills.

现状景观元素
SITE EXISTING ELEMENTS

● 现状树　　　　Existing Trees
　 原有农舍地基　Base of Farmers' House
　 原有农田区域　Farmland Area
　 现状山林区域　Woodland Area

现状山林使场地具备了独特的休憩和生态的功能
Existing woodland provides opportunities for unique recreational & ecological function.

现状坡度
SITE EXISTING SLOPE

> 8%
2-8%
1-2%
< 1%

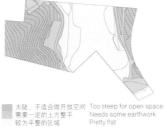

太陡，不适合做开放空间　Too steep for open space
需要一定的土方整平　　　Needs some earthwork
较为平整的区域　　　　　Pretty flat

生态性　　参与性
ECOLOGICAL　PARTICIPATORY

参与性与生态性景观的结合
The integration of participatory and ecological landscape

◁ 场地分析 | Site Analysis
▷ 总平面 | Master Plan

136 / 静谧与欢悦 Tranquility & Joy

01	入口广场	Entry Plaza	13	雨水花园 A	Rain Garden A
02	镜面水池	Interactive Shallow Pool	14	阿基米德花园	Archimedes Garden
03	锈钢板水景墙	Corten Steel Water Wall	15	亲水木平台	Lakeside Resting Area
04	台地坡道	ADA Ramp with Seats	16	湖中小桥	Bridge
05	亲水平台	Sightseeing Plaza	17	活动草坪	Activity Lawn
06	木质座椅	Ex. Tree Planter	18	雨水花园 B	Rain Garden B
07	生态湖	Retention Pond	19	弧形木平台	Resting Area
08	透水混凝土园路	Permeable Concrete Path	20	儿童活动区	Playground
09	活动场地	Sports Field	21	木剧场	Wood Carpet
10	咖啡厅	Cafe	22	攀爬墙	Climbing Wall
11	有机农场	Green House	23	林中小径	Forest Path
12	社区菜园	Kitchen Garden	24	大蚂蚁互动雕塑	Giant Ant Sculpture

山水间 Hillside Eco-Park / 137

雨洪管理系统 | Storm Water Management System

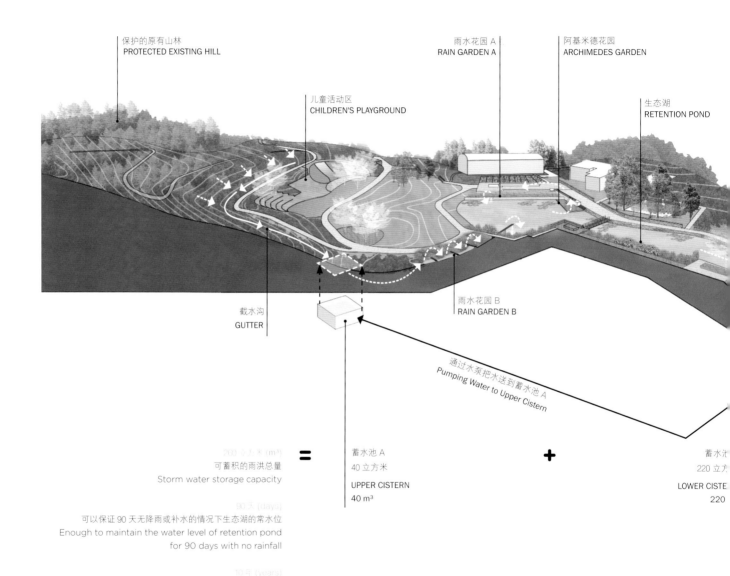

保护的原有山林 PROTECTED EXISTING HILL
儿童活动区 CHILDREN'S PLAYGROUND
雨水花园 A RAIN GARDEN A
阿基米德花园 ARCHIMEDES GARDEN
生态湖 RETENTION POND
截水沟 GUTTER
雨水花园 B RAIN GARDEN B
通过水泵把水送到蓄水池 A Pumping Water to Upper Cistern

260 立方米 (m³)
可蓄积的雨洪总量
Storm water storage capacity

= 蓄水池 A
40 立方米
UPPER CISTERN
40 m³

+ 蓄水池 B
220 立方米
LOWER CISTERN
220 m³

90 天 (days)
可以保证 90 天无降雨或补水的情况下生态湖的常水位
Enough to maintain the water level of retention pond for 90 days with no rainfall

10 年 (years)
可以滞留 10 年一遇的 24 小时连续降雨
Capable of detaining storm flow of 10 year/24 hour storm event

入口广场雨洪管理系统 | Storm Water Management System at Entry Plaza

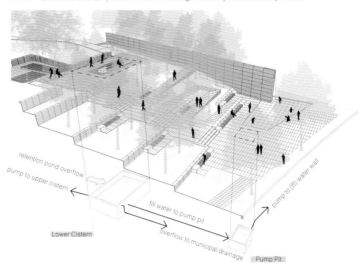

雨水花园系统 | Rain Garden System

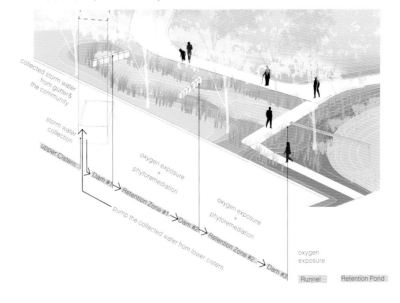

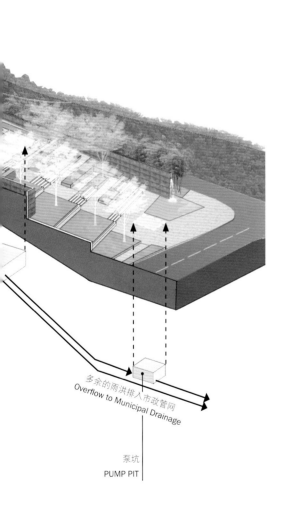

山水间 Hillside Eco-Park / 139

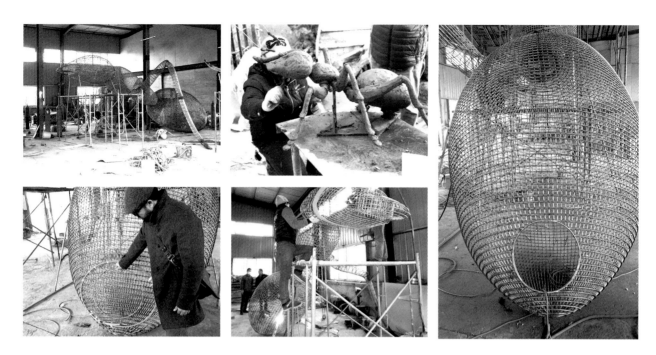

施工过程 | Construction Process

山水间 Hillside Eco-Park / 141

◁ 入口台地 | Entry Terrace
△ 草图研究 | Sketch

山水间 Hillside Eco-Park / 143

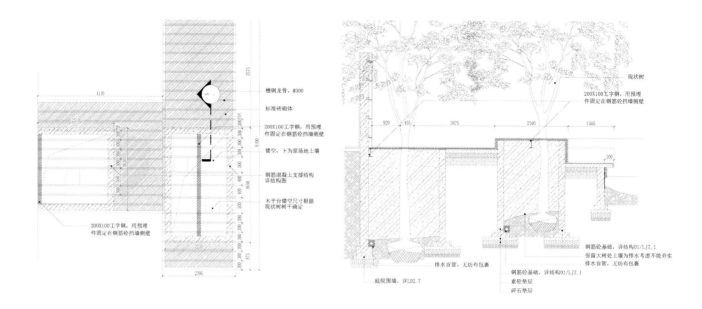

△ 木质座凳施工图 | Construction Drawings of Tree Planter
▽ 施工过程 | Construction Process
▷ 亲水平台 | Sightseeing Plaza

◁ 生态湖 | Retention Pond
△ 雨水花园 | Rain Garden
△ 草图研究 | Sketch

山水间 Hillside Eco-Park / 149

阿基米德花园 | Archimedes Garden

山水间 Hillside Eco-Park / 153

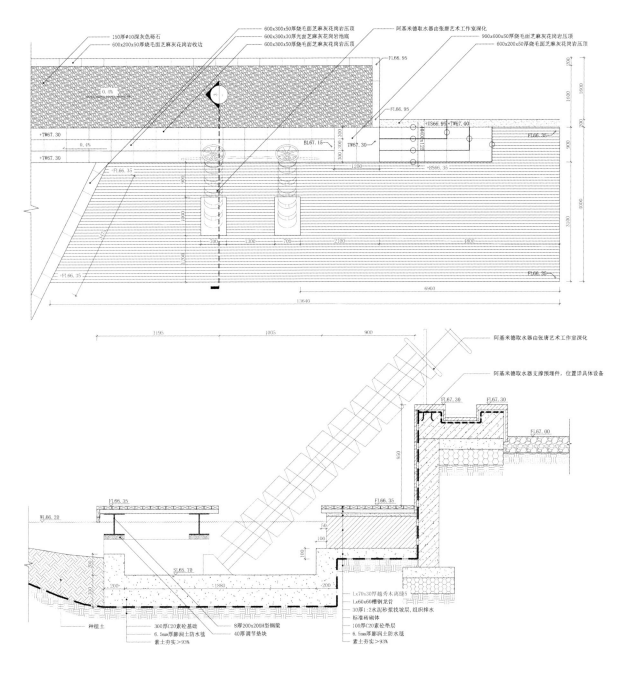

△ 阿基米德花园施工图 | Construction Drawings of Archimedes Garden
▷ 阿基米德花园 | Archimedes Garden

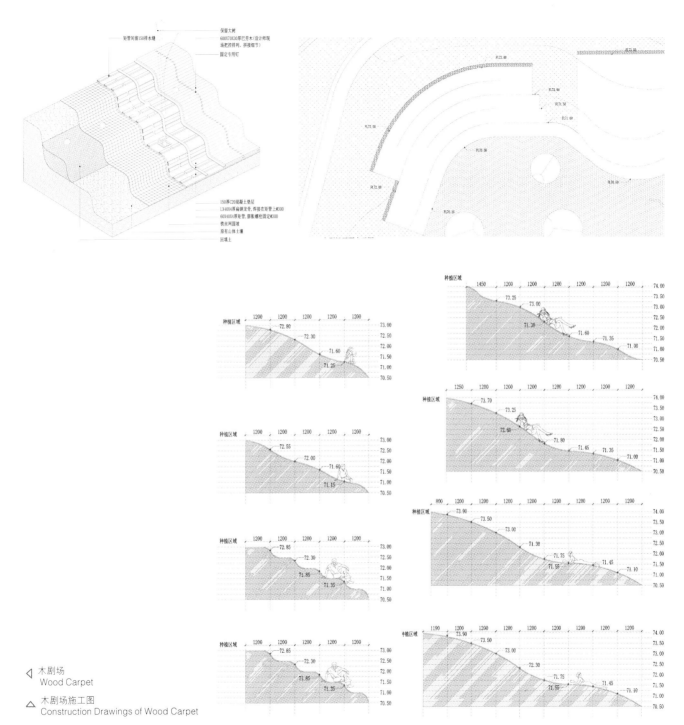

◁ 木剧场
　Wood Carpet

△ 木剧场施工图
　Construction Drawings of Wood Carpet

山水间 Hillside Eco-Park / 157

活动设施 | Activity Facilities

158 / 静谧与欢悦 Tranquility & Joy

良渚文化村
Liangzhu Village

　　良渚文化村是一个城市郊区大型综合性开发项目，位于杭州市西北部，既紧靠著名的良渚文化遗址，又有丘陵绿地和水网平原相结合的自然环境。自2009年以来，张唐景观陆续参与了住宅、商业、公园及酒店内庭等多个项目。

　　春漫里商业广场位于社区中心，服务于四周的商铺、住宅及会所。长达33米的镜面倒影无边水池，成为钟塔在广场上的延伸。竖直挺拔的榉树定义了广场的休息空间。广场尽端的旱喷小广场为节日庆典增加了活跃气氛；周围商、住两用的建筑群内部有5个独立的小庭院，满足居民休息、放松的使用需求，其规范化的管理保证了使用的私密性；滨河公园为市政滨河绿地，规范要求使用尽量少的硬质场地。设计利用相邻住宅工地多余土方营造丰富的地形，塑造起伏变化的空间，采用低维护的本土植物，设计低洼湿地形成场地的自然雨水系统，降低建造和后期维护费用。该公园经过几年的植被恢复，现已成为空间丰富、景致宜人的成熟的滨河绿地。

The Liangzhu Village is a large community in the suburb of northwestern Hangzhou. Since 2009, Z+T Studio has been continuously involved in numerous projects including residential projects, commercial facilities, community parks and hospitality projects.

One of the projects is a community plaza surrounded by commercial facilities, a community center and the residential buildings. The design features a 33-meter long reflecting pool. The borderless pool becomes an extension of the clock tower that sits at one end of the plaza. An array of tall, upward-branching Zelkova trees were planted in the open space, which contributes to the welcoming and relaxing atmosphere of the plaza. At one end of the plaza are several dry deck fountains, which can animate the plaza during special events.

Z+T Studio also designed a series of inner courtyards which provide privacy and safety to the residents while most of the space on the ground floor is for commercial use.

Along the river through the village, Z+T Studio designed a promenade park. It minimized the hard pavement and utilized the soils excavated from the adjacent construction sites to form unique topography. The design of the park also utilized various eco-friendly and low maintenance techniques such as introducing local low maintenance plants, natural rain water collecting which feeds an irrigating system, and more.

榉树广场 | Zelkova Tree Square

△ 旱喷广场 | Fountain Plaza
◁ 水台 | Water Table

滨河草坡 | Riverside Grassy Slope

嘉都公园
Jiadu Park

嘉都是一个距离北京市 30 千米的大型综合社区，中央公园位于四个新建的高层居住组团之间，是一个长 600 米、平均宽度约 55 米的带状社区公园。场地平整，无任何保留植被。基于中国的社会特征，郊区大型社区的入住者大都是有小孩的年轻家庭。对于成年居民来说，公园是他们日常繁忙生活之余放松休闲的地方，也是认识新邻居、建立新的邻里关系的地方；对于在这个社区成长起来的孩子，社区公园会成为他们成长中重要的组成部分。

嘉都中央公园的设计充分考虑到社区公园的使用人群、时间、活动喜好以及气候特征，将公园分为三个主要部分。最西侧的宇宙探索乐园为儿童活动区，以宇宙、星球、虫洞等为主题，成为一个独特的具有寓教于乐作用的活动区；中部生态自然区利用竖向设计汇聚雨水形成人工湿地，湿地两旁设置系列休憩小空间以及适合烧烤和野餐等家庭活动的场地；公园东侧为运动区，包括五人制足球场、篮球场、乒乓球场、单双杠、滑板及多功能活动场地。

该公园是张唐景观对于城市社区公园的一次积极探索和尝试。对社区文化建设、满足日常生活需求、融合生态环境教育和为下一代提供健康成长环境四个方面的思考反映在公园的每一个细节设计上。

Jiadu is a new large residential community located in a suburb of Beijing. A rectangular site at the size of 600-meter by 55-meter surrounded by four high-rise residential building parcels was designated for the Jiadu Central Park. The site is flat and bare. The demographic statistic has indicated that in such a residential community, most residents are young couples with children. For this specific type of community, a park would be the best to provide an essential environment for the residents to relax and meet their new neighbors. For the children, it is crucial to have a safe, healthy and inspiring environment. The community park provides such an environment and encourages the children to take an active part in their communities from an early age.

The design of the Jiadu Central Park took into consideration four key factors — the user groups, the timing, the users' habit and preference and the climate. The park is divided into three zones. At the west end, there is a zone mainly for children's use and themed with the exploration of the universe. In this zone, while playing in the specifically designed landscape, the children can learn about the universe. The zone in the middle is an eco-park centered on a wetland that harvests the runoff rainwater. A series of areas for the family to picnic, barbeque and relax are designed alongside the wetland. The zone at the east end is mainly for sports. It includes various courts for futsal soccer, basketball, table tennis and gymnastics.

Z+T Studio was exploring the opportunities for an urban community park for this project. From the overall design to the details, the Jiadu Central Park has considered — how to encourage the culture of community; how to meet the users' ever-changing requirement in the daily life; how to create an ecofriendly and sustainable environment; and how to provide a healthy, safe and inspiring environment for the children in the community.

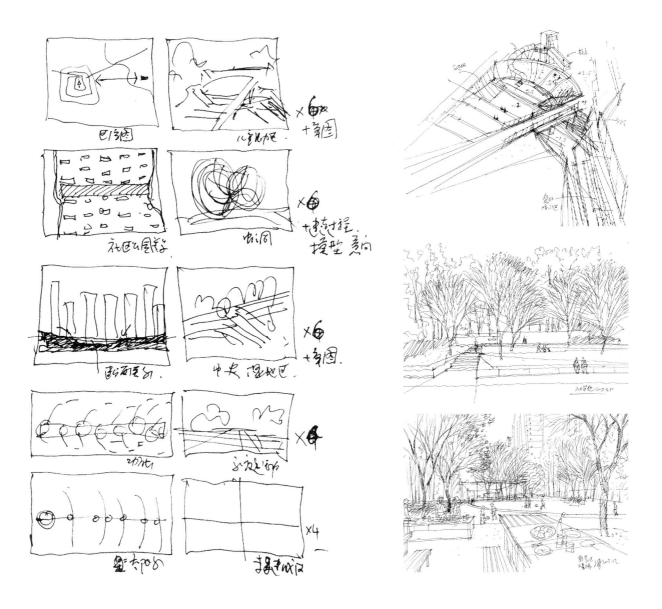

△ 草图研究 | Sketches
◁ 场地现状 | Existing Condition

嘉都公园 Jiadu Park / 177

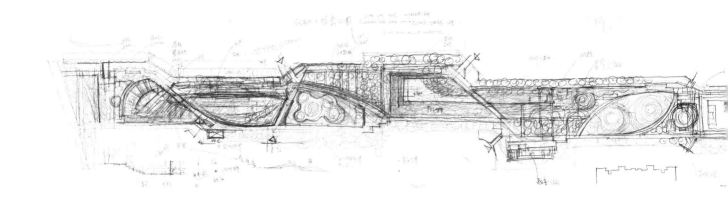

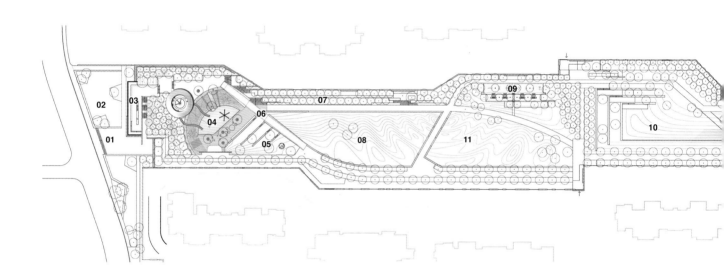

01 西入口 West Entrance	05 草坡 Grass Terrace	09 林下棋牌 Chess Garden	13 五人制足球场 Five-a-side Soccer Field
02 疏林广场 Grove Plaza	06 人行天桥 Pedestrian Bridge	10 雨水花园 Rain Garden	14 乒乓场地 Table Tennis Field
03 眺望台 Gazebo	07 台地花园 Terrace Garden	11 虫洞雕塑 Wormhole Sculpture	15 篮球树练习场 Basketball Tree Court
04 宇宙乐园 Cosmos Paradise	08 多功能大草坪 Multifunctional Lawn	12 活力滑板场 Skateboard Garden	16 复合训练场 Compound Taining Ground

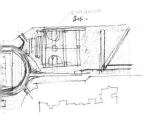

◁ 草图平面 | Sketch

▽ 总平面 | Master Plan

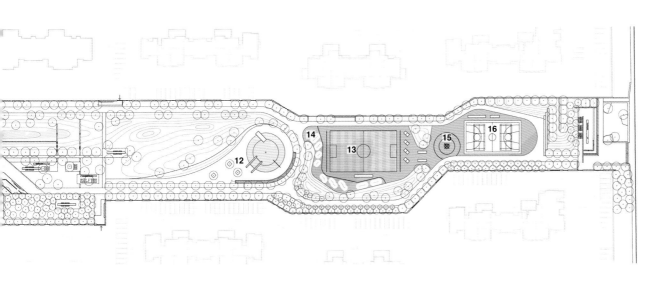

嘉都公园 Jiadu Park / 179

施工过程 | Construction Process

嘉都公园 Jiadu Park

观赏看台 | Terrace

雨水花园 | Rain Garden

引力波草坪 | Wave Lawn with Sculpture

篮球树练习场 | Basketball Court

嘉都公园 Jiadu Park / 189

活动设施 | Activity Facilities

嘉都公园 Jiadu Park

鲸奇谷
Marvel Valley

鲸奇谷位于浙江安吉桃花源度假区的核心区，临近商业街、悦榕庄度假酒店等服务设施。项目用地处于群山间的谷地，有良好的自然植被和丰富的雨水资源。作为一个自然环境中的儿童公园，鲸奇谷充分尊重现有水体、地形和山野的氛围，将各类活动场地和设施巧妙地融入大自然中。放大的蚂蚁窝、山林中的蜂巢、蜘蛛网、水母状戏水池，还有巨大的鲸鱼形的湖面，这些与山体、树林、水体结合的独特的活动场地，让小朋友从不同寻常的视角看世界，发挥他们的想象力。各种主题的活动设施及场地抓住了小朋友最基本的玩乐方式：滑、爬、钻、跳和跑。经典的活动设施滑梯、爬网、秋千、蹦床在鲸奇谷都有它们新的形象和新的玩法，鼓励小朋友在自然中玩耍，搭建他们和自然的联系。鲸奇谷为城市化了的孩子提供了一个平时得不到的环境：玩水、玩沙、玩泥巴。

该项目的特点是在山野的自然环境中植入现代儿童玩乐的方式。设计在充分考虑到安全性的同时，提供各种有趣的活动机会，鼓励小朋友去探索和冒险。由于设计内容和现场山体地形紧密相关，不管在设计阶段还是建设阶段，设计师都需要根据现场的情况反复斟酌，与施工单位配合调整设计，以便达到最佳的效果。

The Marvel Valley is a children's park located at the core of the Taohuayuan Resorts in Anji County, Zhejiang Province. The Marvel Valley resides in a basin formed by adjoining hills. The basin offers very well preserved vegetation and serves as a natural rain water collector. To preserve the existing natural conditions, the park is designed to utilize the existing hilly topography, tree groves, the vegetation on the hills and the large pond. A series of special playgrounds were designed for the park with the theme of an ant's nest, beehives, spider nets, octopus and whale. Each playground features uniquely designed equipment for specific activities. These programs reflect the natural characteristics of the site, allowing the landscape design to integrate the playgrounds into the nature. The park offers unique perspectives and vantage points for the children to inspire their imagination. The playground equipment and the sites are designed to assist children develop their essential motor skills and physical behavior. In the Marvel Valley, the classic games like slides, crawling net, swing and trampoline have been given a new look, and appear as fresh and unique experiences for the children. The design encourages city children to establish a rare connection and relationship with nature by allowing them to interact playfully with the most basic natural elements such as water, sand and earth.

Having fully considered the safety requirement, the design of the programs also encourages exploration and adventure. In order to seamlessly integrate with the existing topography, whether during the design or construction phases, the landscape architects repeatedly adjusted and fine-tuned the design to achieve the best possible result.

场地现状 | Existing Condition

草图研究 | Sketch

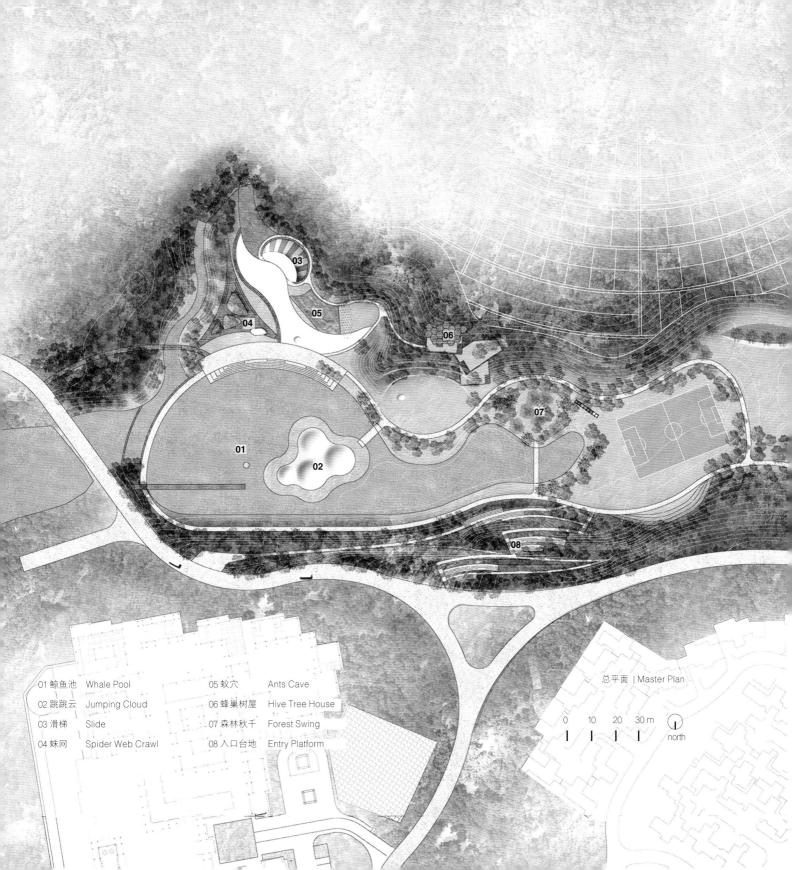

198 / 静谧与欢悦 Tranquility & Joy

施工过程 | Construction Process

入口台地 | Entry Terrace

△ 蜂巢树屋 | Hive Tree House
◁ 板岩小径 | Slate Path

水磨石滑梯 | Terrazzo Slide

鯨奇谷 Marvel Valley / 209

水母戏水池 | Jellyfish Padding Pools

鯨奇谷 Marvel Valley / 211

植物细节 | Plant Details

CMP 广场
CMP Plaza

该项目位于青岛，是一个包括办公、商业、公寓、酒店等功能的城市近郊综合性开发项目。景观设计结合园区规划框架，对街道景观、入口广场以及各个组团提出不同的形象定位。从园区主题为切入点，以"映像之城、像素、棱光、光圈"等主题定位各个区块，并力求以景观设计语汇加以诠释。从而强调各个街区的独特风格，增强区块的识别性和领域感，提供新鲜独特的景观感受。

在探索形式的同时，雨水的处理方式被纳入整个设计系统，尝试在传统雨水收集系统上增加雨水花园，以缓解日趋严重的城市雨水系统压力。材料方面也不断探索新型环保材料，大量利用预制混凝土（PC）、透水混凝土等环保材料，以体现园区的主要宗旨。

该项目是一次对城市路口街角空间有效利用的积极探索。西北角的主入口广场景观将园区与城市融为一体，利用高差设计浅水面喷泉水景，兼顾水景效果和放空后的广场实用功能。建成后成为附近居民休闲聚集的场所，使城市规划中的不同界面有了良好的渗透和互动。

The CMP Plaza is a mixed-use project which includes office, commercial, residential and hotel zones. Z+T Studio was commissioned to design the entry of the community near a major traffic intersection. The following building features were considered mandatory — a part of the city streetscape, an entry plaza of the main building and the entrance to the residential compounds. Z+T Studio conceptually divided the entire space into several parts, each with its own individual theme — reflection of the city, pixel, prism and aperture respectively. The spaces are distinguished by the different themes rather than by physical partitions. The strategy serves the goals of the project while preserving the continuity of the urban space.

A system to collect and recycle rain water was also introduced to the project. It includes a rain garden that recycles the collected runoff. Combining the collecting and recycling features, the system is a relief to the municipal drainage system and the high demands placed on municipal irrigation water. The design also features sustainable and eco-friendly material such as precast and porous concrete.

Since the main plaza at the northwest corner of the site is intended as an integrated part of the cityscape, the design utilizes site elevation difference to create a shallow water pool with fountains. It functions as a major water feature and an open plaza when the water is drained. It has become a welcoming gathering place for residents in the neighborhood. The plaza has connected the project to the urban space at large.

倒影池 | Reflecting Pool

倒影池 | Reflecting Pool

云朵乐园
Cloud Paradise

云朵乐园是成都麓湖生态城内道路和湖面之间的一个狭长的滨水绿地，面积约25 000平方米。由于是在现有条件上进行改造，设计有许多限制条件，包括已有湖岸线、码头、紧急消防通道等。受到麓湖生态城人造湖水系统的启发，云朵乐园的主要概念是将公园儿童活动功能和对水的环境教育功能结合，形成一个寓教于乐的公园。它既是一个有趣的儿童公园，又是一个露天的自然博物馆。水的各种形态及汇集形式，云、雨、冰、雪，以及溪流、河道、池塘、漩涡等都被巧妙地结合在活动场地和节点设计中，形成跳跳云、互动旱喷广场、曲溪流欢、涌泉戏水池、冰川峡谷镜面墙、雪坡滑梯、漩涡爬网这些独特的活动场地。

设计师深度参与建造过程是该项目能顺利进行的一个重要因素。张唐景观艺术工作室整合设计师、艺术家、结构师、工程师和技术工人，从设计和建造上对一些重要节点深度把握，在工厂内部测试、制作，然后现场安装。既能够有效降低造价、控制工期，又能保证质量确保设计的意图能很好地实现。

云朵乐园建成后在行业内、公众中反响良好。正式开放前，管理方邀请十几个家庭进行内部测试，并据此对设计、细节、管理和维护进行了优化调整。为了确保每一个人都能获得良好的体验，公园对每一天的人流量进行了一定限制。

The Cloud Paradise is a 25,000 ㎡ promenade along the side of the lake in the Luhu Eco City in Chengdu, Sichuan Province. The design was tightly restricted by existing conditions including the preservation of the existing structures, the contour of the lake and the fire code. Inspired by the design of the water system of the Luhu Eco City, water is the theme in the Cloud Paradise. The Cloud Paradise was designed as a children's park meant to teach about water and the natural environment. It was designed to not only serve as a playful park, but also an outdoor natural museum. Z+T Studio designed the bouncing cloud, the interactive dry deck fountain plaza, the creeks, the interactive spring water pond, the glacier mirror wall valley, the snowy slides and the whirlpool crawling net. Through these uniquely designed landscape features and program, information related to water forms like clouds, rain, creeks, rivers, ponds, ice, snow and whirlpools is presented in special ways.

In order to achieve the design goals, the designers at Z+T Studio participated heavily in the construction phase. As the consortium of designers, artists, structural engineers, and technicians, the Art Workshop of Z+T Studio played a key role from the design to the final realization of the project. The Art Workshop designed and fabricated many key elements. Prior to the final production and installation, the Art Workshop constructed the full-size mock-up and ran intensive tests of the design in the Z+T Studio's workshop space. Through the participation of the Art Workshop and this unique design-built procedure, the cost of the production, the schedule, the quality and the design integrity have been better controlled compared to conventional procedures.

Prior to its grand opening, the management of the park invited a dozen families to test play the facilities in the park. Based on their feedback, together with Z+T Studio, the management fine-tuned the details and optimized maintenance strategies. For a better experience in the park, the management also established a ceiling for the overall number of daily visitor. Since its opening, the Cloud Paradise has gained positive feedback from the public and professionals.

◁ 场地现状 | Existing Condition
▷ 草图研究 | Sketches

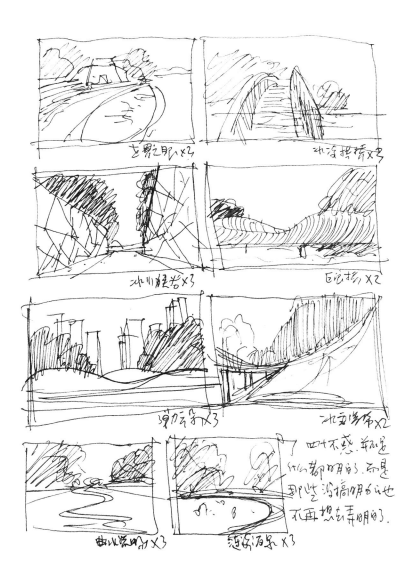

云朵乐园 Cloud Paradise / 223

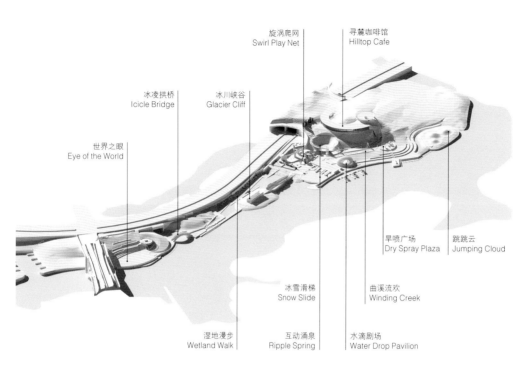
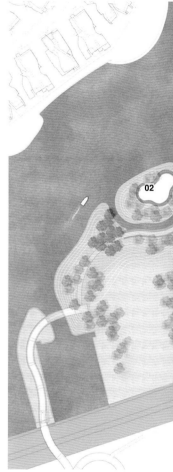

△ 轴测图解 | Axonometric Diagram
▷ 总平面 | Master Plan

01 曲水流欢 Winding Creek
02 跳跳云 Jumping Cloud
03 旱喷广场 Dry Spray Plaza
04 互动涌泉 Interactive Fountain
05 冰雪滑梯 Ice Slide
06 旋涡爬网 Climbing Net
07 水滴剧场 Water Droplet Theater
08 湿地漫步 Wetland Garden
09 巨浪飞渡 Wave Bridge
10 冰川峡谷 Glacier Canyon
11 冰凌拱桥 Ice Bridge
12 世界之眼 Eye of the World
13 寻麓咖啡厅 Hilltop Cafe
14 小卖部 Kiosk
15 码头 Wharf

世界之眼 | Eye of the World

世界之眼细节 | Eye of the World Detail

旱喷广场 | Fountain Plaza

曲溪流欢 | Winding Creek

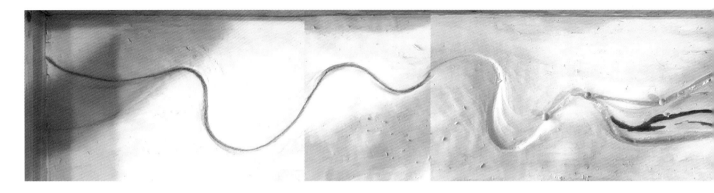

△ 曲溪流欢模型 | Winding Creek Model
▽ 施工过程 | Construction Process

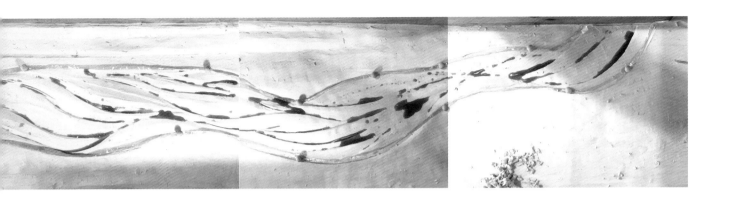

云朵乐园 Cloud Paradise / 235

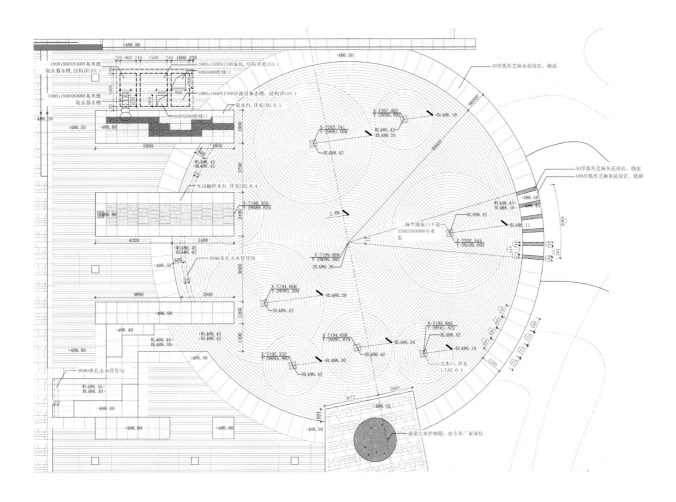
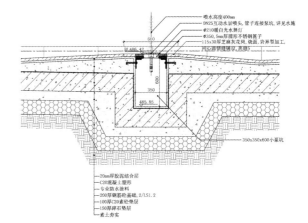

△ 互动涌泉及施工图 | Interactive Fountain and Its Construction Drawings
◁ 互动涌泉鸟瞰 | Interactive Fountain Bird's Eye View

云朵乐园 Cloud Paradise / 239

◁ 冰川峡谷现场 | Glacier Canyon Existing Condition
▽ 草图研究 | Sketch
▷ 施工过程 | Construction Process

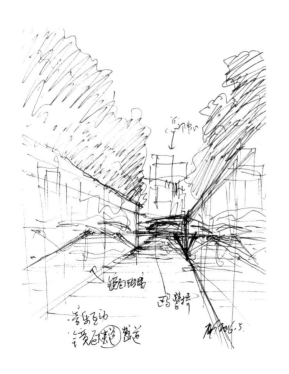

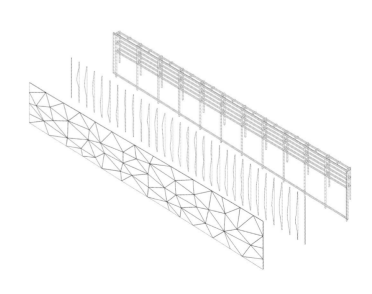

242 / 静谧与欢悦 Tranquility & Joy

云朵乐园 Cloud Paradise / 243

水滴剧场 | Water Droplet Theater

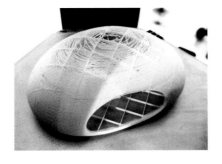  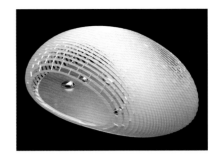

△ 水滴剧场电脑模型 | Digital Model of Water Droplet Theater
▽ 施工过程 | Construction Process

云朵乐园 Cloud Paradise / 245

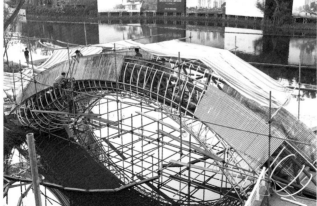

◁ 冰凌拱桥电脑模型和施工图
Digital Model of the Ice Bridge & Construction Process

▷ 冰凌拱桥 | The Ice Bridge

互动涌泉 | Interactive Fountain

冰雪滑梯 | Ice Slide

活动设施 | Activity Facilities

公园里
The Park

公园里位于苏州吴江区新开发区域一个路口的两侧，周边住宅区密布但缺乏公共活动空间。场地被一条城市道路分为东、西两部分。东侧的现状建筑围合出场地的基本空间形态；西侧街角的一块为永久性开放空间，剩下的部分为未来商业用地。根据项目的特殊位置和现状条件，我们将场地划分为东、西广场和临时小公园三个部分，希望能赋予场地更多的城市公共空间属性，使其成为一个具有包容性和开放性的场所，欢迎周边社区的居民前来休闲利用。

东、西两个广场均以缓坡和城市人行道联系，统一设计的铺装模糊了场地和市政人行道空间的界限，赋予场地友好的对外界面。广场上几组大型抬高种植池对空间进一步围合划分。东、西广场上的两组水景设计独具特色。东广场利用建筑和广场之间的高差设计波浪水台，造浪机推出的一阵阵水浪和花岗岩雕刻的石材共同形成丰富而细腻的亲水景观。西广场在地面上利用现状高差设计浅水海浪戏水池，鼓励小朋友参与其中。临时小公园主要由起伏的草坡地形和可再次利用的活动设施组成。器械和场地以自然中的花、果、叶、昆虫为主题，给孩子们营造了一个童话般的微观世界。

The Wujiang New Development Zone is a newly established urban district in Suzhou, Jiangsu Province. While the district has quickly filled with high density residential projects, there are few public spaces. The Park is a project to provide such needed public areas. The site is divided into east and west by a municipal street. On the east side, existing buildings define the available space for the Park; on the west side, the space was further divided into two sections — the part close to the street was designated as a permanent public space, and the part further west would eventually be for commercial use but temporarily is designated for public use. Under these circumstances, the design solution was to create two public plazas with more permanent elements and a park with more transitional landscape features. The goal of the project is to transform the site to an inclusive, open, welcoming and playful public space especially for the residents in the neighborhood.

The design utilizes the gradual slopes and pavement similar to the municipal pedestrian sidewalk system, blurring the boundary between the two. A series of elevated planters defines the spaces within the plazas. Two unique water features on the east and west plazas give the Park its distinctive characteristics. On the east plaza, taking advantage of the elevation difference between the building and the plaza, Z+T Studio designed two water wave platforms. A pair of specially designed machines embedded in the platforms create water waves simulating the movements of the wave on the beach. The waves, spray, sound and relief made from granite on the platforms creates a unique and animated scene. On the west plaza, also taking advantage of the elevation difference on the site, Z+T Studio designed an accessible shallow pool with an embedded wave generator. The interactive scene in the middle of the urban area is unique and has become a major attraction for the children in the neighborhood. For the temporary park, Z+T Studio utilized the soil on the site to reform the topography, which allows for easier rearrangements in the future. On the new topography, the low rolling hills, grass slopes and lawns were created. Movable playground equipment was arranged to create several themed playgrounds. The hilly topography and nature-themed playgrounds have created unique and fairy tale-like places for the children.

◁ 场地现状 | Existing Condition
▷ 草图研究 | Sketch
▽ 总平面 | Master Plan

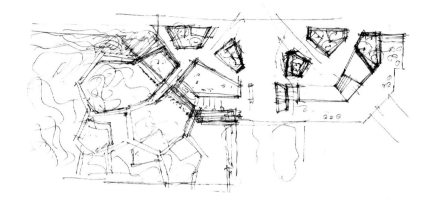

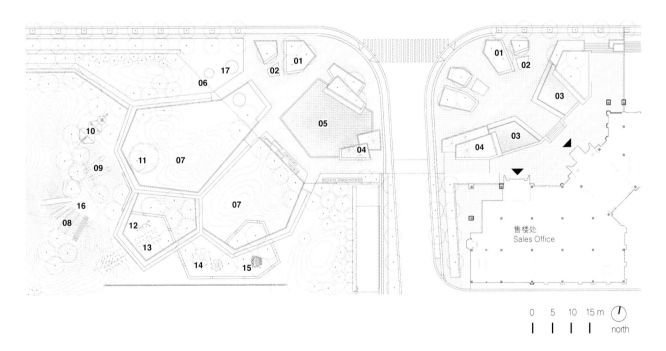

01 种植池座凳 Seating Planter
02 爵士白石凳 Granite Seat
03 波浪水台 Wave Water Table
04 蒲公英雕塑 Dandelion Sculpture
05 造浪戏水池 Wave Pool
06 安全围栏 Railing
07 草坡地形 Grass Slope
08 大滑梯 Giant Slide
09 钻洞 Climbing Tunnel
10 树叶攀爬网 Leaf Climbing Net
11 秋千 Swing
12 大青虫传声筒 Caterpillar Megaphone
13 音乐迷宫 Music Maze
14 花瓣跳跳板 Flower Jumper
15 莲蓬雕塑 Lotus Sculpture
16 沙坑 Sand Pit

静谧与欢悦 Tranquility & Joy

施工过程 | Construction Process

公园里 The Park / 257

波浪水台 | Wave Water Table

公园里 The Park / 261

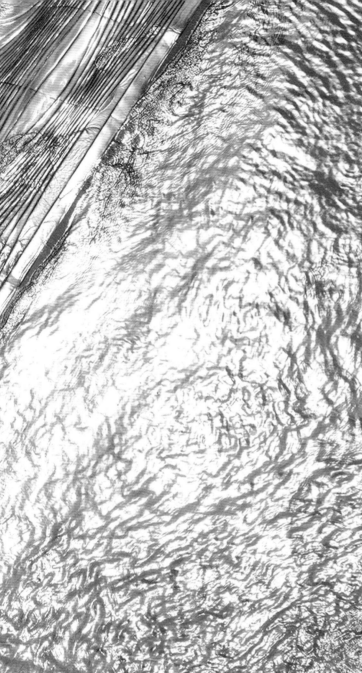

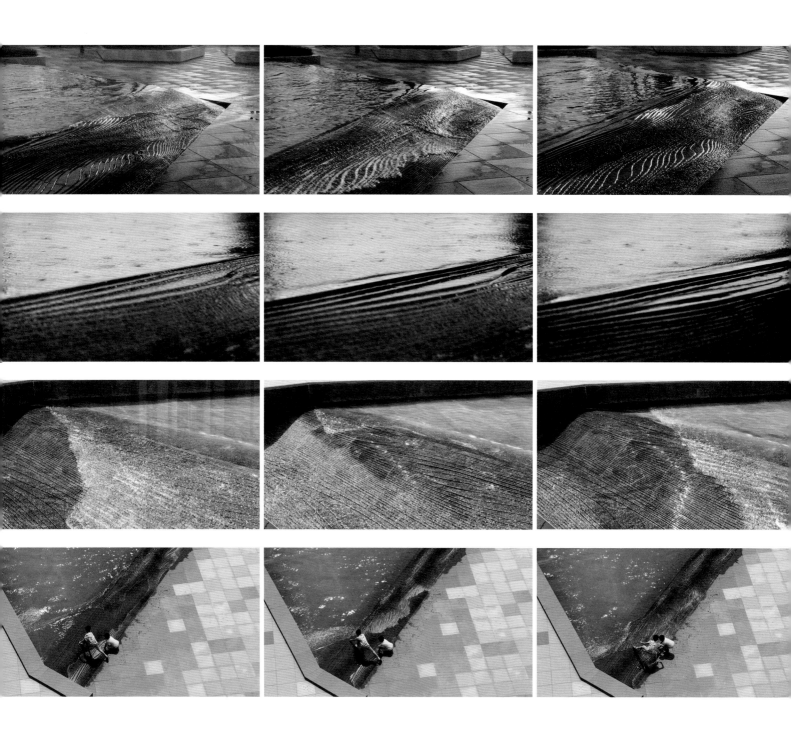

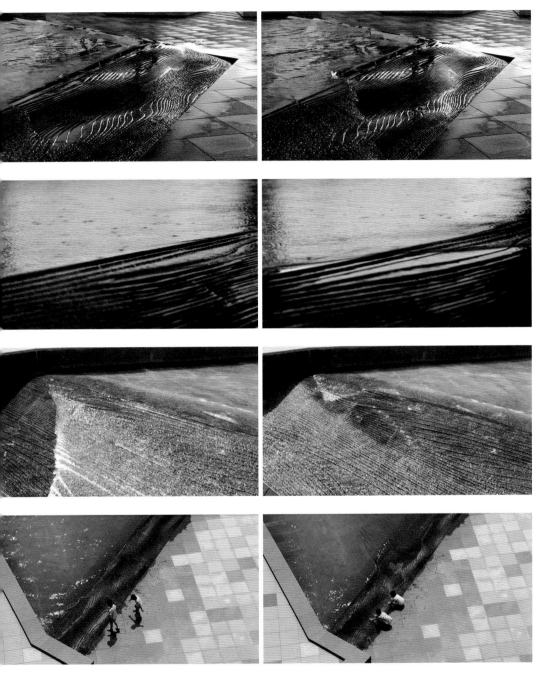

波浪水台分帧 | Wave Water Table Video Stills

公园里 The Park / 265

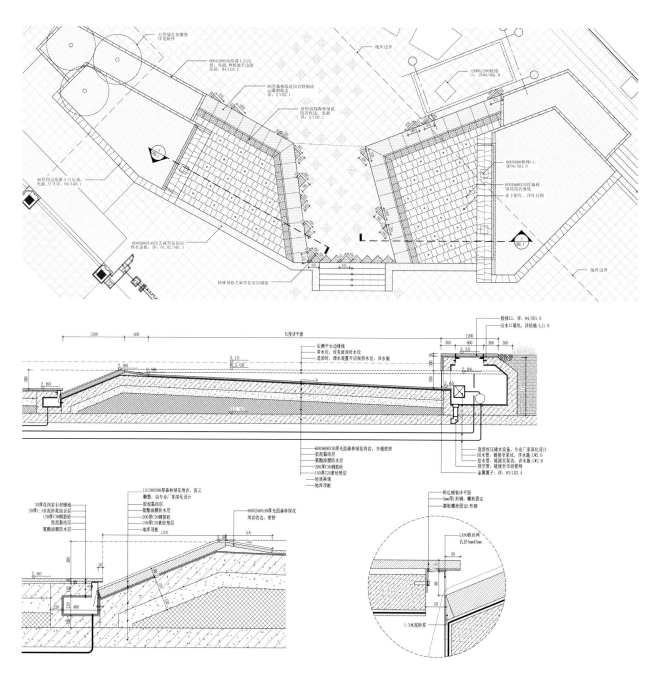

△ 波浪水台施工图 | Wave Water Table Construction Details
▷ 波浪水台细节 | Wave Water Table Details

造浪戏水池 | Wave Pool

造浪戏水池 | Wave Pool

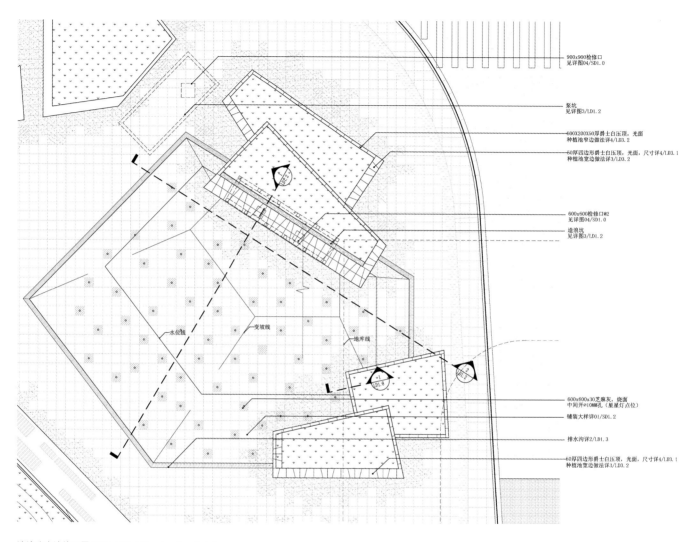

造浪戏水池施工图 | Wave Pool Construction Details

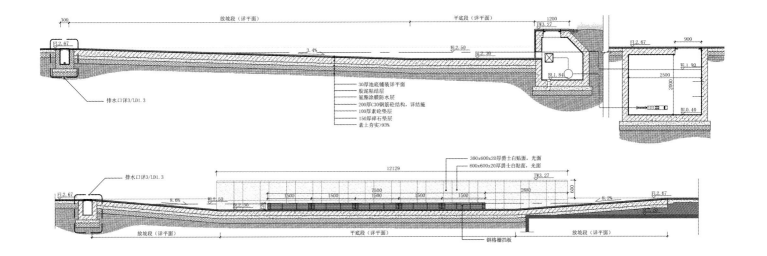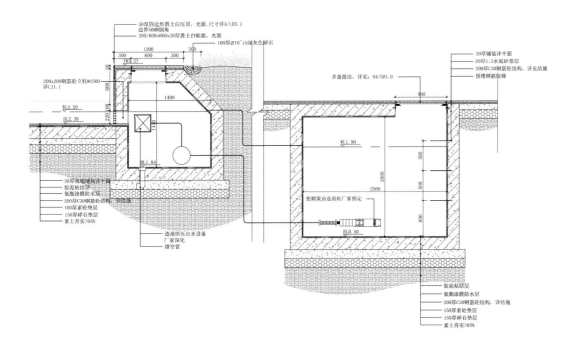

活动设施 | Activity Facilities

理想城
Dream City

理想城位于上海北部郊区，定位为现代风格、小家庭的酒店式公寓（loft）生活社区。景观面积约为3.3公顷。分为临街商业、下沉商业街，以及南、北两个组团空间。景观总体设计采用形式感强的多边形切分场地，在满足基本功能的同时，刻意强调了整体的俯瞰效果，为周围高层建筑上的住户提供独特的视觉体验；在商业区进入下沉广场的扶梯一侧设计了特色水景，希望将人群从地面引导至地下商业，下沉广场的地面与顶部利用环形座凳、照明等系列装饰以及丰富的色彩增加其商业气氛。

水景设计以不同的水景形态搭配、组合来丰富空间感受。通过在材料及节点设计上进行试验和创新，综合利用玻璃、金属、石材的不同质感和特性，营造出与众不同的景观感受。项目中第一次尝试使用玻璃作为跌水水景的材料，其中对跌水效果、安装、维护等设计过程的考虑，对提升整体商业区域的氛围，起到了积极的作用。

The Dream City is a modern residential complex located in the north of Shanghai. The complex emphasizes the modern loft-living life style. The 3.3-hectares outdoor space includes the street-level commercial buildings, a sunken commercial plaza, and north and south courtyards. The landscape design uses geometrical aesthetic principles to divide the site into multiple areas. Taking advantage of the sunken plaza, the design created vivid graphical views for the vantage points from the street level and the surrounding high-rise buildings. A waterfall feature made from glass plates is the highlight connecting the street-level commercial spaces to the sunken plaza.

The design of waterscape brings diversified experience to the space with different combinations of water features. A variety of types of glasses, metals, and stones are used. For the first time, glass is used as a structural material to create falling water features in the project. With the consideration of water-falling effects, tectonic and maintenance, the waterscape plays a dynamic role in enriching the atmosphere and the overall experience of the space.

商业内庭 | Commercial Chamber

东方传奇
Oriental Legend

西安万科东方传奇示范区位于西安市极具文化底蕴的曲江新区，在满足景观功能的基础上完成文化特征的植入成为本项目的重要课题。设计方案探讨了现代"中式"的可能性，以王维两首脍炙人口的诗句作为概念展开了设计。"行到水穷处，坐看云起时"为广场区的概念，根据场地形态和竖向特征，方案设计了"镜池""拱泉""平山""对影""流云"及"壁道"几个各具特征的节点；"明月松间照，清泉石上流"为内街的概念，方案以一系列水景串联起了"石溪""双瀑""莲池""曲溪"及"清潭"几个节点。方案还引入了大量不同种类的大、中型乔木以及观赏草类，在营造亲切自然的氛围的同时，与几处水景一起，在商业街内形成了舒适的小气候。

The Vanke Oriental Legend is located at the Qujiang New District in Xi'an, Shaanxi Province. The district has a rich cultural history. The challenge of the design is to integrate the cultural context into the project using the content of the landscape design program. The design explores new possibilities within the modern Chinese style. Traditional Chinese poems become the inspiration for the design. The main plaza was inspired by the lines in a famous poem by Wang Wei, a Tang Dynasty poet, "Go to the place where there is scarcity of water, sit and watch the clouds rise." While working with the actual site conditions, the design introduced a series of nodes — the reflecting pool, arch fountains, miniature mountains, pairing shadows and the flowing clouds. For the inner street includes more lines of poetry by Wang Wei, "The bright moon shines between the pines, the clear spring water flows over the stones"are the inspiration. Several spaces are interconnected through a series of focal points such as the stone creek, the pair of the falling water, the lotus lake, the meandering creek and the crystal-clear pond. Throughout the project, the design has also introduced a variety of plants including large trees and ornamental grasses. The diversified plants and series of water features together create a comfortable microclimate in the space.

△ 平山地形 | Miniature Mountains Landform
◁ 镜池旱喷 | Arch Fountain Plaza

曲溪石台 | Meandering Stone Creek

流云叠瀑 | Flowing Clouds Sculpture

优盛广场
U-Center Plaza

项目位于北京海淀区五道口城铁站旁，坐落于优盛大厦东侧，是一处介于建筑与自行车棚之间的狭长硬质空地。业主计划将该空地优化为配套商业的城市开放空间，并期望通过策略性设计，改善人群消费体验、营建具备活力的标志性景观。基于现状条件及综合分析，本案设计最大的挑战是如何激活缺乏活力的城市中建筑的边角剩余空间，并将其打造成为具有独特魅力的城市开放空间。

项目紧邻一条有五十多年历史的火车线路，每当火车经过时，所有车流人流均停下来等待。受到这个场地特殊的城市体验的启发，我们提出了主题概念："等待下一个十分钟"。在场地中安装一个转盘，上面的喷泉、座椅、树与场地的一排旱喷、一排树成一系列。而尽头的一组喷泉和树在圆盘里，可以转动。转动持续50分钟，当转盘里的这组泉和树回到初始的位置时，场地内所有的泉水开始涌动。喷水持续10分钟。然后继续下一个循环。一个小时，是我们设定的一个时间的度量器，让人在有限的空间里感受一段时间的流逝。该主题的引入力图让使用者在有限空间中获得感受时光流逝且循环往复的宇宙观，同时凸显场地特性。

景观设计引入的转盘，在工业领域有着成熟的技术和设备。它使得该城市的更新项目在有限的预算内成功地激活了场地，使其成为对每一个人都欢迎和包容的城市地标性场所。

The U-Center Plaza is a pedestrian promenade located near the Wudaokou Railway Crossing in Haidian District, Beijing. The project was meant to revitalize a linear space between the U-Center and a bicycle parking pavilion. The space had been used as a shortcut passageway and a temporary parking lot. The client had hoped the project could transfer this back-of-the-house space to a dynamic promenade while also offering the public a positive and pleasant experience and give the U-Center a new identity. Based on the existing conditions and the requirement, the largest challenge was how to turn a forgotten place into an open and dynamic urban plaza.

The project is adjacent to a 50-year old railway crossing which, when passing, forces all traffic to stop for a certain amount of time. Inspired by such a common experience, the project is designed with a special theme — Waiting for the Next 10 Minutes. Z+T Studio designed a 17-meter diameter turntable as the core of the promenade's hardscape. As the turntable rotates, its position is linked to the on-and-off of the fountains and lights embedded in the shallow pool across the promenade. One section of the shallow pool is embedded in the turntable. Every 50 minutes when the turntable completes one rotation of 360-degree, both sections of the shallow pool will be aligned. The alignment triggers the fountains and lights, turning the promenade to a vibrant festival walk. The celebration lasts 10 minutes while the movement of the turntable recesses. Pedestrians on the promenade are often caught by surprise by the sudden spray of fountains while others are waiting patiently anticipating this moment. As the cycle simulates the movement pattern at the railway crossing, the project materializes and transforms the concept of time to an experience. This dynamic feature has become the identity of the space.

It has set a successful example for how to grow the landscape design vocabulary while continuing to revitalize the urban environment. The incorporation of modern technologies and machinery is not only to make the design more viable to construct, but also, more importantly, to transform the position of the landscape from passive to proactive, from a presenter to a participant. Through the design that involves the understanding of the context and the innovative technology, the landscape architecture goes beyond a showpiece and becomes a generator of a new layered experience and sense of space that is part of the ever-evolving urban life.

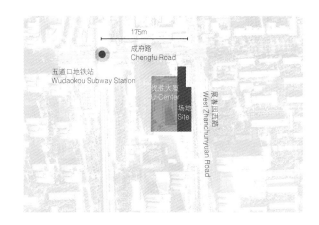

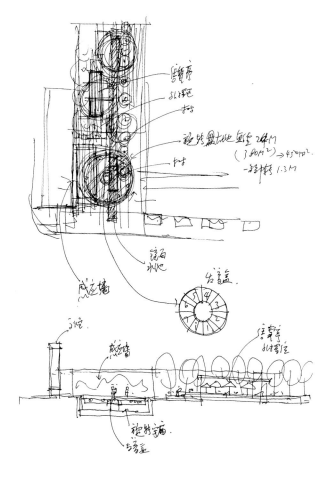

◁ 场地现状 | Existing Condition
△ 项目区位 | Project Location
▷ 草图研究 | Sketches

总平面 | Master Plan

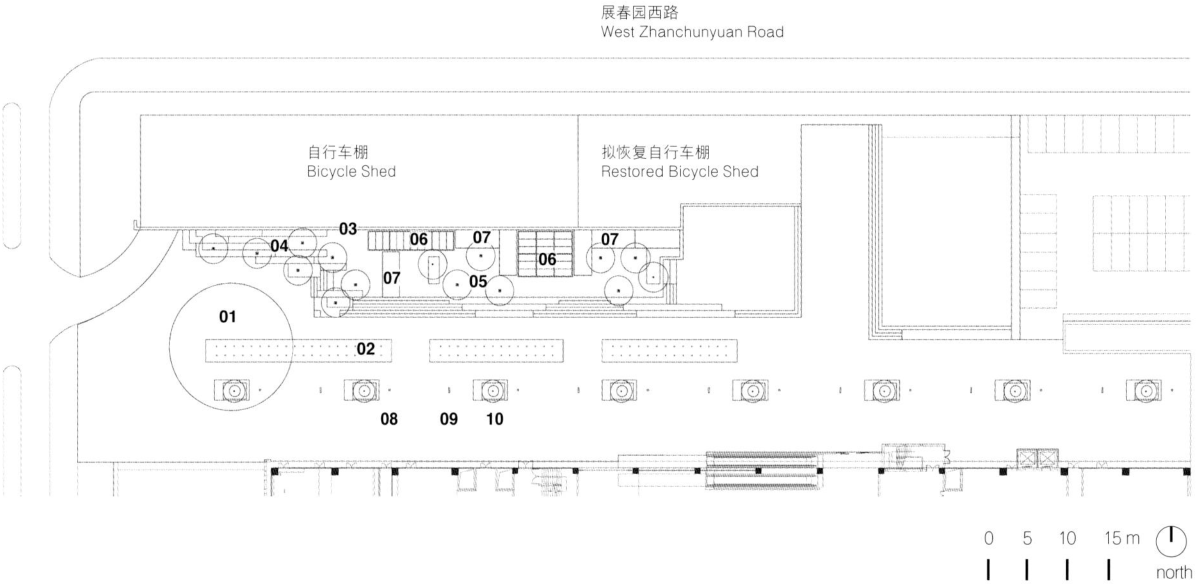

展春园西路
West Zhanchunyuan Road
自行车棚
Bicycle Shed
拟恢复自行车棚
Restored Bicycle Shed
01
02
03
04
05
06
06
07
07
07
08
09
10
0 5 10 15 m
north

01 旋转平台 Rotating Platform
02 音乐喷泉 Music Fountain
03 涂鸦墙 Graffitti Wall
04 广场看台 Amphitheater
05 休闲台地 Coffee Terrace
06 廊架 Pergola
07 集装箱售货亭 Container Kiosk
08 灯柱 Lighting Pole
09 广告牌 Billboard
10 种植池座凳 Seating Planter

轴测图 | Axonometric Drawing

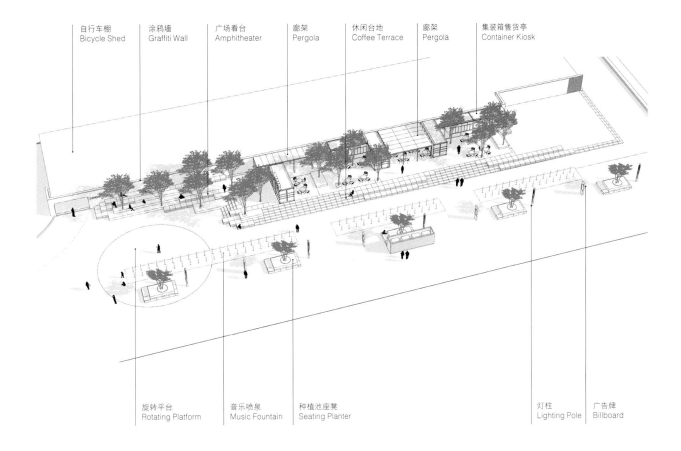

自行车棚 Bicycle Shed
涂鸦墙 Graffiti Wall
广场看台 Amphitheater
廊架 Pergola
休闲台地 Coffee Terrace
廊架 Pergola
集装箱售货亭 Container Kiosk

旋转平台 Rotating Platform
音乐喷泉 Music Fountain
种植池座凳 Seating Planter
灯柱 Lighting Pole
广告牌 Billboard

优盛广场 U-Center Plaza / 295

施工过程 | Construction Process

优盛广场 U-Center Plaza / 297

旋转平台 | Rotating Platform

298 / 静谧与欢悦 Tranquility & Joy

优盛广场 U-Center Plaza

300 / 静谧与欢悦 Tranquility & Joy

◁ 旋转平台分帧 | Rotating Platform Video Stills
▽ 旋转平台图解 | Rotating Platform Diagram

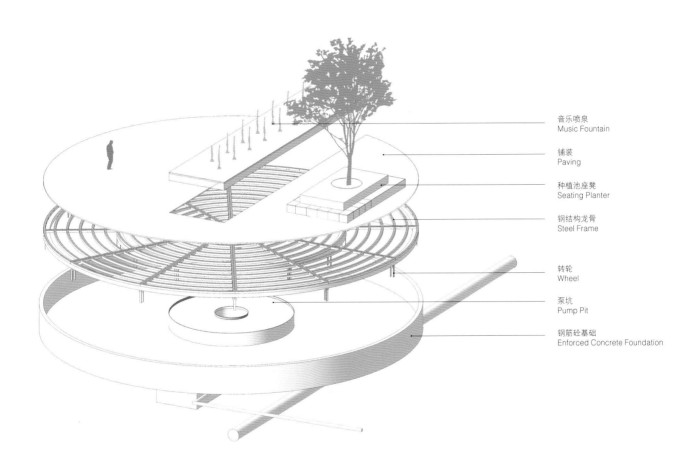

音乐喷泉
Music Fountain

铺装
Paving

种植池座凳
Seating Planter

钢结构龙骨
Steel Frame

转轮
Wheel

泵坑
Pump Pit

钢筋砼基础
Enforced Concrete Foundation

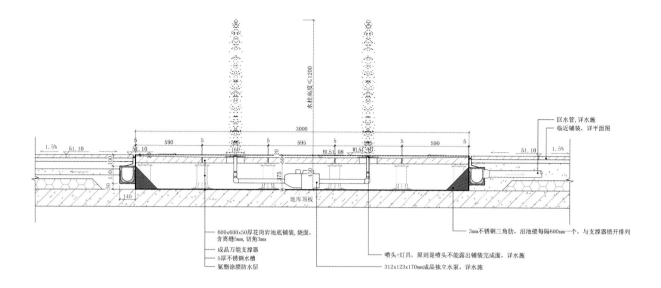

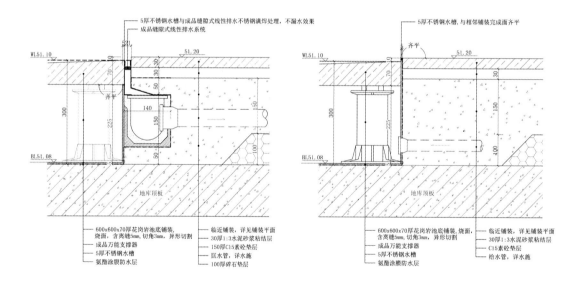

△ 音乐喷泉施工图 | Music Fountain Construction Details
▷ 音乐喷泉细节 | Music Fountain Details

音乐喷泉细节 | Music Fountain Details

广场看台 | Wooden Terrace

附录

公司简介

张东先生和唐子颖女士于2009年在上海共同创立了上海张唐景观设计事务所（简称"张唐景观"）。经过近年来在景观规划、场地设计、艺术装置以及生态技术等方面的实践，张唐景观在全球景观设计界的影响力逐渐扩大，获得了包括美国景观设计师协会（ASLA）荣誉奖在内的多个奖项。张唐景观认为，自然与艺术为主导要素的景观设计对人们生活的幸福感有着直接的影响。通过对"参与性景观"设计理念的实践，张唐景观倡导一种具备环境教育功能、可提升场所活力并增进人与人的关系，同时又符合场地特征的独特设计手段。

张东和唐子颖在北美和中国有着丰富的学习及工作经历，这为张唐景观奠定了理解东、西方不同思维方式的基础，也因此在工作中更加强调多元化及多学科间的合作。在中国设计市场飞速变化、大型设计机构不断涌现的当下，张唐景观始终刻意保持一种相对较小的规模，以确保首席设计师有足够的时间和精力投入到每个项目当中。通过事务所的集体智慧、创新文化及多元化团队成员的通力合作提升项目品质。

张唐景观认为，任何好的设计都离不开大胆的构想和深入的研究。因此，张唐景观创立了艺术工作室（Art Workshop）以及自然工作室（Biophilic Lab）。艺术工作室通过对创新材料及构造方法的研究，使张唐景观的设计能力得到了显著提升，同时通过全尺寸模型的制作、测试和生产，确保交付产品的质量；自然工作室则致力于推进张唐景观"参与性生态"的景观实践工作，在场地设计中积极应对来自水环境、能源、栖息地及物种多样性等议题日益严峻的挑战，从而提升设计作品的科学性和可持续性。艺术工作室和自然工作室在不同程度上支撑着景观项目的创新性及可持续性。

Appendix

Company Profile

Z+T Studio was founded by Mr. Dong Zhang and Ms. Ziying Tang in Shanghai in 2009. Since then, the studio has gained solid and critical reputation for its design, diverse works and unique design approach. Z+T Studio's portfolio includes master planning, research centers, public parks and city furniture. Z+T Studio has received numerous national and international awards, including the ASLA General Design Honor Awards.

Situated at the cross-road of the landscape architecture itself and in the increasingly urbanized environment, Z+T understands the profession has evolved beyond the traditional focus on aesthetics and gardens, becoming a catalyst and engine to encompass and enrich the area's diversity, self-esteem and self-identity. With a strong emphasize on the most basic elements of nature, Z+T has used its novel approach to ratify its core values —— to connect nature to community well-being and strengthening landscape resilience and social ecology sustainability.

Prior to founding Z+T Studio in Shanghai, Mr. Zhang and Ms. Tang had extensive academic and practice experience in the U.S. and China. Such an experience laid the foundation for the firm to embrace the perspectives and sensitivities of both West and East.

Z+T Studio is a consortium in a landscape design atelier, an Art Workshop and a Biophilic Lab with design professionals, artist, fabricators and horticultural specialists and ecologists. The unique structure and multi-disciplinary collaboration not only enables Z+T to take on sophisticated projects, but also to transform the firm into an engine to initiate innovative and cutting-edge design ideas. Even while accepting complex and large-scale projects, the two founding principals of Z+T have maintained an intimately-sized atelier and hands-on working method for their commitment to each project.

艺术工作室
Art Workshop

自 2014 年成立以来，艺术工作室拓展了张唐景观的设计范畴，提升了景观设计作品的品质，同时也为张唐景观在跨学科合作的尝试提供了一种新的思路。

艺术工作室团队由艺术家、设计师、制造商、工程师和承包商共同组成。工作室集成了电脑模拟与 3D 打印技术、现代机械和声光电技术，拥有能够加工各种材料的能力。作为张唐景观的重要组成部分，工作室在最初的概念设计、草模推敲到生成成品这一过程中，始终同景观设计事务所紧密配合，同时也提供建造的可行性、绩效分析、维护和材料等方面的信息，为景观设计的决策过程提供重要参考。

艺术工作室的设置模糊了各专业间的边界，它不仅仅提供了技术上的支持，更是一种灵感获取的有效途径。这种相互交织的关系促使张唐景观的设计具有文化敏感性，能够促进人们对场地的参与和感知。

In 2014, Z+T Studio founded the Art Workshop. Since then, the workshop not only has expanded Z+T's design and execution capability, but also empowered Z+T's vision of the intimate, yet multi-disciplinary collaboration in landscape design.

The Art Workshop includes artists, designers, fabricators, engineers and contractors. The workshop is equipped with computer molding and 3D-printing technology, acousto-optic technology and machinery that is capable of dealing with a wide range of materials as well as a large factory-like workspace. As an integral part of Z+T Studio, the workshop works with the design atelier from initiating to developing the design ideas and elements through study models and typologies to producing the sophisticated final products in the workspace and on-site. The workshop also offers information on and evaluations of manufacturing feasibility, cost, maintenance and materiality that are critical for design decisions.

自然工作室
Biophilic Lab

　　无论项目规模的大小，景观设计本身都有着生态保护和修复的责任。张唐景观创建自然工作室，将园艺、土木工程和生态学的知识进行整合，并融入项目的全设计过程当中。自然工作室的融入有助于降低新建项目对环境的不利影响，并提升改造项目场地恢复的能力。

　　自然工作室致力于推进张唐景观"参与性生态"的景观实践工作，同时也是可持续性建设评估及设计策略的顾问机构。为了应对日益严峻的生态环境挑战，自然工作室通过亲自然设计，不断追求场地"活力"与"生态"之间更高品质的平衡关系。

As landscape design has evolved beyond mere aesthetics to protecting and healing the ecosystem at all scales, Z+T Studio has integrated a Biophilic Lab. The lab brings the expertise of the horticultural specialists and ecologists to the design atelier. Throughout the entire design process, the lab helps to design ecologically sound systems that improves the ecological well-being of the site while decreasing the environmental impact to the site by the new project.

Since one of Z+T's core initiatives is "Participatory Ecology", the Biophilic Lab is dedicated to sustainable and innovative research and design from a biophilic perspective. Through biophilic design, the project is not only environmentally sound, but also addresses the human reconnection to the natural world and vice versa.

设计理念

景观设计是艺术,是源于对大自然的热爱和理解,基于场地需求,对自然元素提炼、抽象和重组的艺术。

我们倡导源于自然的设计,遵循自然的原始生态原则:可持续发展的生态设计是后工业时代景观的必经之路。源于自然的设计,遵循的是生态原则的本质。通过对现代生态技术的应用,使新开发的场地得到最佳的生态发展;高密度开发或者已经遭到破坏的自然景观,得到恢复和健康开发。

我们倡导基于场地的设计,寻求独特的场所精神:对场地人文、地理、生态和项目需求的全面和深入的分析,是项目独创性的源泉。寻找场地最为重要的基本元素将其提炼、升华和抽象,赋予场地灵魂;通过创造性的手法营造有强烈的可辨认性的现代景观场所。

我们倡导对自然元素的提炼、抽象和重组,创造人与自然的再联系:人对自然的认知是本能的。可感知的自然、便于记忆的场所,以及能辨认的城市景观给人们归属感。

现代工业文明制造了人与自然之间的障碍。"五色令人目盲,五音令人耳聋",越来越多的人工建成的居住和生活场所或者缺乏设计,或者被强加太多的设计,隔离了人对阳光、空气、水以及自然生命的直接感知。

我们探索自然的复杂性与统一性,简单和复杂之间的转换和平衡,单纯景观元素在外界条件改变下的无穷变化和可能性,以及复杂要素之间单纯的逻辑关系。遵循自然的原始生态原则和场地精神,通过对自然元素的提炼、抽象和重组,重塑人与自然的联系。

张唐景观认为:
- 景观设计是提升人们幸福感的一种途径。
- 景观设计传递我们对生活的感悟和对美好生活的向往。
- 关注生活本身:多样化的空间、场地和设施提高生活丰富度。
- 简约设计:把握对阳光、气象、水、植物等自然元素的直观审美。
- 尺度宜人,用材精简,工艺合理,提供持久的审美感受。

Philosophy

In Z+T Studio, landscape design is meant to refine nature based on a comprehension of artistic principle in nature. It extracts and reintegrates landscape elements based on the individual character of each site.

Refined nature, respect original eco-system: Respecting the natural eco-system is a living Chinese philosophy. Existing ecology needs to be closely studied before any transformation is considered. By applying modern eco-techniques, the site will be optimally developed. High density development or a contaminated environment needs to be regenerated in a healthy direction.

Restored eco-system, has specific site spirit: All sites are different despite the presence of modern industrialized development. Any innovation should be inspired by the specific elements (weather, wind, light and shade) of site. The recovered site will be identified by the created spatial organization and design.

Reformed site, recall the memory of site: Industrialization create obstacles between people and their environment. Non-designed or overdesigned spaces are full of constructed, human-made places. We look through to the essence of each site while also exploring the relationship between complexity and simplicity within nature. At all time we seek to reconnect people with water, sunlight, and plants through our use of refined landscape elements.

We see landscape design as:
- a connector of nature and community well-being.
- a vessel for our vision of sustainability and resilience.
- a catalyst and engine to encompass and enrich diversity, self-esteem and self-identity.
- a distilling process towards the essence and the most basic elements of nature.

团队
Team

张东 Dong Zhang	唐子颖 Ziying Tang	张玫芳 Meifang Zhang	赵桦 Hua Zhao	张亚男 Yanan Zhang	徐敏 Min Xu
2009 —	2009 —	2010 —	2010 —	2009 —	2012 —
杜强 Qiang Du	董万荣 Wanrong Dong	陈逸帆 Yifan Chen	姚瑜 Yu Yao	袁帅 Shuai Yuan	孙川 Chuan Sun
2009 — 2017	2009 — 2011	2015 —	2012 —	2016 —	2015 —
吴叶飞 Yefei Wu	钱沁禾 Qinhe Qian	周士奇 Shiqi Zhou	顾欣骏 Xinjun Gu	王虎 Hu Wang	牛宇轩 Yuxuan Niu
2010 — 2011	2017 —	2017 —	2016 —	2016 —	2016 —
张晓珏 Xiaojue Zhang	范景飞 Jingfei Fan	姜雪婷 Xueting Jiang	张喆鑫 Zhexin Zhang	侯天宇 Tianyu Hou	喻咏菲 Yongfei Yu
2009 — 2011	2012 — 2013	2013 — 2014	2018 —	2015 — 2016	2015 — 2016

墨 Mo Wang 15 —	周啸 Xiao Zhou 2013 —	张卿 Qing Zhang 2011 —	范炎杰 YanJie Fan 2009 —	刘洪超 Hongchao Liu 2011 —	郑佳林 Jialin Zheng 2011 —
少豪 Shaohao Bian 16 —	蔡孙喜 Sunxi Cai 2011 — 2016	王琪 Qi Wang 2017 —	刘昕 Xin Liu 2013 — 2018	薛阿男 Anan Xue 2009 — 2014	林佩勋 Peixun Lin 2013 — 2016
姝晗 Shuhan Qin 11 — 2016	杨玉鹏 Yupeng Yang 2017 —	彭阳 Peng Yang 2015 — 2018	席琦 Qi Xi 2016 — 2018	张明富 Mingfu Zhang 2011 — 2012	杜欣波 Xinbo Du 2017 —
劼睿 Jierui Wei 17 — 2018	王晨 Chen Wang 2012 — 2016	李进妹 Jinmei Li 2011 —	胡一昊 Yihao Hu 2016 — 2018	陈宇 Yu Chen 2009 —	秦静 Jing Qin 2018 —

项目信息

樾园
苏州 / 2016 / 980 ㎡
团队：张东、唐子颖、杜强、范炎杰、郑佳林、刘洪超、孙川、林佩勳、张玫芳、喻咏菲、侯天宇、姚瑜、张卿、秦姝晗、袁帅、牛宇轩
业主：中航里城有限公司
奖项：ASLA 2017年专业奖（公共设计类荣誉奖）
发表文章："时光雕刻——苏州航樾园内庭设计"，《风景园林》，2016.12；"Yue-yuan Courtyard"墨西哥 LANDUUM 杂志，2017.05.

永泰会所
上海 / 2010 / 10 000 ㎡
团队：张东、唐子颖、董万荣、张亚男、范炎杰、张晓珏、杜强
业主：江苏东元湾联合投资有限公司
发表文章："景观设计的实现"，domus，2011.12.

玖著里
宁波 / 2016 / 5 000 ㎡
团队：张东、唐子颖、徐敏、姚瑜、赵桦、卞少豪、袁帅、杨玉鹏、关天愉
业主：宁波万科
发表文章："古典的隐喻——宁波玖著里景观设计思考"，《中国园林》2017.08；"古典的隐喻—宁波玖著里"《风景园林》2017.08.

富力十号
杭州 / 2016 / 5 580 ㎡
团队：张东、唐子颖、张亚男、张玫芳、林佩勳、郑佳林、刘洪超、孙川、王晨
业主：杭州富力

九里云松
杭州 / 2013 / 6 850 ㎡
团队：张东、唐子颖、张亚男、杜强、吴叶飞、薛阿男、范炎杰、张卿、郑佳林
业主：九里云松度假酒店开发有限公司
发表文章："中式极简的道与术——杭州九里云松度假酒店的改造与重生"，《中国园林》，2015.11；"Pins De La Brume Hotel"，Landscape Record 杂志，2016.10.

玉湖会所
昆山 / 2013 / 8 310 ㎡
团队：张东、唐子颖、赵桦、徐敏、秦姝晗、张玫芳、杜强、刘洪超、薛阿男、刘昕、范景飞、姜雪婷
业主：昆山中节能环保投资有限公司

建研中心
东莞 / 2012 / 18 500 ㎡
团队：张东、唐子颖、杜强、赵桦、董万荣、张晓珏、张玫芳、范炎杰、秦姝晗
业主：万科建筑技术研究有限公司
奖项：ASLA 2014年专业奖（通用设计类荣誉奖）
发表文章："生态景观技术与艺术探索——广东省东莞市万科建研中心生态园区"，《景观设计学》，2014.03；"Vanke Research Center"，韩国 Landscape World 杂志，2015,Vol.78；"Vanke Research Center"，《建筑知识》，2015.05；"The SMART Landscape"，Images Publishing Group,2016.04；"Vanke Research Center"，爱沙尼亚 moodnekodu 杂志，2016.06.

京华园
北京 / 2009 / 3 600 ㎡
团队：张东、唐子颖、董万荣、杜强、张晓珏
业主：北京天卉苑花卉研究所
奖项：北京市设计金奖，2009

山水间
长沙 / 2016 / 14 000 ㎡
团队：张东、唐子颖、周啸、张亚男、赵桦、郑佳林、刘洪超、蔡孙喜、刘昕、姜雪婷、王墨、姚瑜、林佩勳、陈逸帆、张卿、秦姝晗、王晨
业主：中航里城有限公司
发表文章："长沙中航城国际社区"山水间"公园"，《风景园林》，2015.07；"长沙中航城国际社区"山水间"公园的参与性与生态性"，《景观设计学》，2016.02；"The Hillside Eco-Park"，意大利 PAYSAGE 杂志，2017.01.

良渚文化村
杭州 / 2012 / 80 320 ㎡
团队：张东、唐子颖、张亚男、杜强、范炎杰、张晓珏、张明富、董万荣、刘洪超、蔡孙喜
业主：杭州余杭新都工程开发有限公司

嘉都公园
北京 / 2017 / 36 000 ㎡
团队：张东、唐子颖、张卿、周啸、徐敏、范炎杰、刘洪超、郑佳林、胡一昊、王墨、蔡孙喜、顾欣骏、牛宇轩、魏劼睿
业主：北京佰嘉置业集团

鲸奇谷
安吉 / 2017 / 30 000 ㎡
团队：唐子颖、张东、张亚男、徐敏、陈逸帆、刘洪超、郑佳林、胡一昊、孙川、刘昕、顾欣骏、杜强、彭阳、牛宇轩、卞少豪、王琪
业主：浙江绿城元和房地产开发有限公司
发表文章："与儿童互动的景观设计"，《旅游规划与设计》，22 期；"Marvel Valley"，意大利 PAYSAGE 杂志，2018.02.

CMP 广场
青岛 / 2013 / 6 500 ㎡
团队：张东、唐子颖、赵桦、张卿、张玫芳、秦姝晗
业主：青岛万科

云朵乐园
成都 / 2017 / 25 000 ㎡
团队：张东、唐子颖、张卿、郑佳林、刘洪超、孙川、胡一昊、徐敏、周啸、彭阳、顾欣骏、席琦、卞少豪、王琪
业主：成都万华新城发展股份有限公司
发表文章："孩子们的自然博物馆——成都麓湖生态城云朵乐园"，《景观设计学》，2017.06；"数字化景观节点的深化设计研究——以成都麓湖云朵乐园为例"，《风景园林》，2017.11.

公园里
苏州 / 2016 / 13 000 ㎡
团队：张东、唐子颖、周啸、徐敏、张卿、顾欣俊、郑佳林、刘洪超、孙川、杜强
业主：苏南万科上海区域万晟产品能力中心
发表文章："设计传递美好——苏州公园里展示区景观设计思考"，《中国园林》，2017.01.

理想城
上海 / 2013 / 3 600 ㎡
团队：张东、唐子颖、赵桦、张卿、张玫芳、杜强、郑佳林
业主：上海致臻实业发展有限公司

东方传奇
西安 / 2015 / 10 000 ㎡
团队：张东、唐子颖、赵桦、周啸、刘昕、张卿、姚瑜、范炎杰、刘洪超、郑佳林、孙川、王墨、林佩勳、秦姝晗
业主：西安万科企业有限公司
发表文章："西安万科东方传奇展示区"，《风景园林》，2015.11.

优盛广场
北京 / 2016 / 3 600 ㎡
团队：张东、唐子颖、赵桦、刘洪超、范炎杰、孙川、郑佳林、徐敏、刘昕
业主：北京邦泰摩尔资产管理有限公司
发表文章："等待下一个十分钟——北京五道口优盛大厦广场改造"，《景观设计学》，2016.6；"WAITING FOR THE NEXT TEN MINUTES:U-CENTER PLAZA"，Landscape Record 杂志，2017.04.

劝学公园	P2～3
云朵乐园	P4～5
山水间	P6～7
东原惠南	P8～9
荷兰花海	P10～11

Project Credits

Yueyuan Courtyard
Suzhou/2016/980 ㎡
Design Team: Dong Zhang, Ziying Tang, Qiang Du, Yanjie Fan, Jialin Zheng, Hongchao Liu, Chuan Sun, Peixun Lin, Meifang Zhang, Yongfei Yu, Tianyu Hou, Yu Yao, Qing Zhang, Shuhan Qin, Shuai Yuan, Yuxuan Niu
Client: Avic Legend Co. Ltd.
Award: 2017 ASLA Professional Award (General Design)
Publication:
1- "Time Sculping—Suzhou Yueyuan Courtyard Design", *Landscape Architecture*, 2016.12.
2- "Yueyuan Courtyard", *LANDUUM*, 2017.05.

Yongtai Club
Shanghai/2010/10,000 ㎡
Design Team: Dong Zhang, Ziying Tang, Wanrong Dong, Yanan Zhang, Yanjie Fan, Xiaojue Zhang, Qiang Du
Client: Jiangsu Dongyuanwan consolidated investment Co., Ltd
Publication: "Implementation of landscape design", *domus*, 2011.12.

Jiu Zhu Li
Ningbo/2016/5,000 ㎡
Design Team: Dong Zhang, Ziying Tang, Min Xu, Yu Yao, Hua Zhao, Shaohao Bian, Shuai Yuan, Yupeng Yang, Tianyu Guan
Client: Ningbo Vanke
Publication:
1- "The Metaphor of classical gardens: Landscape design of jiuzhuli, Ningbo", *Landscape Architecture*, 2017.08.
2- "The Metaphor of classical gardens: Landscape design of jiuzhuli, Ningbo", *Chinese Landscape Architecture*, 2017.08.

Royal Territory
Hangzhou/2016/5,580 ㎡
Design Team: Dong Zhang, Ziying Tang, Yanan Zhang, Meifang Zhang, Peixun Lin, Jialin Zheng, Hongchao Liu, Chuan Sun, Chen Wang
Client: Hangzhou R&F Properties

Pins De La Brume
Hangzhou/2013/6,850 ㎡
Design Team: Dong Zhang, Ziying Tang, Yanan Zhang, Qiang Du, Yefei Wu, Anan Xue, Yanjie Fan, Qing Zhang, Jialin Zheng
Client: Pins De La Brume Hotel, Hangzhou
Publication:
1- "The Paradigm of Chinese Landscape Minimalism- The Rejuvenation of the Pins De La Brume Bouquet Hotel Hangzhou", *Chinese Landscape Architecture*, 2015.11.
2- Pins De La Brume Hotel", *Landscape Record*, 2016.10.

Yuhu Club
Kunshan/2013/8,310 ㎡
Design Team: Dong Zhang, Ziying Tang, Hua Zhao, Min Xu, Shuhan Qin, Meifang Zhang, Qiang Du, Hongchao Liu, Anan Xue, Xin Liu, Jingfei Fan, Xueting Jiang
Client: Kunshanzhong Energy Conservation and Environmental Protection Investment Co., Ltd.

Vanke Research Center
Dongguan/2012/18,500 ㎡
Design Team: Dong Zhang, Ziying Tang, Qiang Du, Hua Zhao, Wanrong Dong, Xiaojue Zhang, Meifang Zhang, Yanjie Fan, Shuhan Qin
Client: Vanke Architecture Technology Co., Ltd.
Award: 2014 ASLA Professional Awards (Honor Award)
Publication:
1- "Art and Eco-Technology, Eco-campus of Vanke Architecture Research Center in Dongguan, Guangdong", *LAF*, 2014.03.
2- "Vanke Research Center", *Landscape World*, 2015, Vol.78.
3- "Vanke Research Center", *Architectural Knowledge* 2015.05.
4- "The SMART Landscape", *Images Publishing Group*, 2016.04.
5- "Vanke Research Center", *moodnekodu*, 2016.06.

Jinghua Garden
Beijing/2009/3,600 ㎡
Design Team: Dong Zhang, Ziying Tang, Wanrong Dong, Qiang Du, Xiaojue Zhang
Client: The 7th China Flower Expo.
Award: Design Award-Gold, 2009

Hillside Eco-Park
Changsha/2016/14,000 ㎡
Design Team: Dong Zhang, Ziying Tang, Xiao Zhou, Yanan Zhang, Hua Zhao, Jialin Zheng, Hongchao Liu, Sunxi Cai, Xin Liu, Xueting Jiang, Mo Wang, Yu Yao, Peixun Lin, Yifan Chen, Qing Zhang, Shuhan Qin, Chen Wang
Client: Avic Legend Co. Ltd.
Publication:
1- "Hillside Eco-Park of Zhonghang Caticity Community in Changsha", *Landscape Architecture*, 2015.07.
2- "Hillside Eco-Park of Changsha Zhonghang Caticity Community- Participatory and Ecological Characteristics of a Typical Chinese Neighborhood Park", *LAF*, 2016.02.
3- "The Hillside Eco-Park", *PAYSAGE*, 2017.01.

Liangzhu Village
Hangzhou/2012/80,320 ㎡
Design Team: Dong Zhang,Ziying Tang,
Yanan Zhang,Qiang Du,Yanjie Fan,
Xiaojue Zhang,Mingfu Zhang,Wanrong Dong,
Hongchao Liu,Sunxi Cai
Client: Hangzhou Vanke
Contractor:Hangzhou Tianqin
Landscape Engineering Co., Ltd.

Jiadu Park
Beijing/2017/36,000 ㎡
Design Team: Dong Zhang,Ziying Tang,
Qing Zhang,Xiao Zhou,Min Xu,
Yanjie Fan,Hongchao Liu,Jialin Zheng,
Yihao Hu,Mo Wang,Sunxi Cai,Xinjun Gu,
Yuxuan Niu,Jierui Wei
Client: Sanhe City Baijia Real
Estate Co., Ltd.

Marvel Valley
Anji/2017/30,000 ㎡
Design Team: Ziying Tang,Dong Zhang,
Yanan Zhang,Min Xu,Yifan Chen,
Hongchao Liu,Jialin Zheng,
Yihao Hu,Chuan Sun,Xin Liu,Xinjun Gu,
Qiang Du,Yang Peng,Yuxuan Niu,
Shaohao Bian,Qi Wang
Client: Greentown Real Estate Co.,Ltd.
Publication:
1- "Interactive Landscape Design
with Children", *Tourism Planning &
Design*，No. 22.
2- "Marvel Valley", *PAYSAGE*,
2018.02.

CMP Plaza
Qingdao/2013/6,500 ㎡
Design Team: Dong Zhang,Ziying Tang,
Hua Zhao,Qing Zhang,Meifang Zhang,
Shuhan Qin
Client:Qingdao Vanke

Cloud Paradise
Chengdu/2017/25,000 ㎡
Design Team: Dong Zhang,Ziying Tang,
Qing Zhang,Jialin Zheng,Hongchao Liu,
Chuan Sun,Yihao Hu,Min Xu,
Xiao Zhou,Yang Peng,Xinjun Gu,
Qi Xi,Shaohao Bian,Qi Wang
Client: Chengdu Wanhua New Town
Development Co.,Ltd.
Publication:
1- "A Natural Museum for Children:
Cloud Paradise in Luxe Lake Eco-city,
Chengdu", *LAF*, 2017.06.
2- "Explore the Digital Design Method
on Landscape Architectural Node-
Taking the Cloud Park in Chengdu
Luxe Lake as an Example", *Landscape
Architecture*,2017.11.

The Park
Suzhou/2016/13,000 ㎡
Design Team: Dong Zhang,Ziying Tang,
Xiao Zhou,Min Xu,Qing Zhang,
Xinjun Gu,Jialin Zheng,Hongchao Liu,
Chuan Sun,Qiang Du
Client: Sunan Vanke (Shanghai Area)
Publication: "Conveying the Insights
and Aspiration of Beauty through
Design- 'The Park' in Suzhou",
Chinese Landscape Architecture，2017.01.

Dream City
Shanghai/2013/3,600 ㎡
Design Team: Dong Zhang,Ziying Tang,
Hua Zhao,Qing Zhang,Meifang Zhang,
Qiang Du,Jialin Zheng
Client: Shanghai Zhizhen Industrial
Development Co., Ltd.

Oriental Legend
Xi'an/2015/10,000 ㎡
Design Team: Dong Zhang,Ziying Tang,
Hua Zhao,Xiao Zhou,Xin Liu,
Qing Zhang,Yu Yao,Yanjie Fan,Hongchao Liu,
Jialin Zheng,Chuan Sun,Mo Wang,
Peixun Lin,Shuhan Qin

Client: Xi'an Vanke
Publication: "Xi'an Vanke Oriental
Legend", *Landscape Architecture*,2017.11.

U-Center Plaza
Beijing/2016/3,600 ㎡
Design Team: Dong Zhang,Ziying Tang,
Hua Zhao,Hongchao Liu,Yanjie Fan,
Chuan Sun,Jialin Zheng,Min Xu,Xin Liu
Client: Bontop Group
Publication:
1- "Waiting For the Next Ten minutes
——Redevelopment of the U-Center
Plaza, Wudaokou, Beijing", *LAF*,
2016.06.
2- "WAITING FOR THE NEXT TEN
MINUTES:U-CENTER PLAZA", *Landscape
Record*,2017.04.

Quanxue Park	P2～3
Cloud Paradise	P4～5
Hillside Eco-Park	P6～7
Venus Mansion	P8～9
Dutch Flower Field	P10～11

摄影 | Photo Credits

张海 | Hai Zhang
2, 4, 6, 8, 10, 13, 23, 32, 34, 36, 39, 42, 44, 46, 49, 50, 51, 52, 53, 55, 62, 63, 65, 66, 67, 68, 70, 72, 73, 75, 76, 77, 78, 79, 81, 90, 91, 92, 94, 96, 98, 100, 101, 103, 104, 106, 107, 108, 109, 111, 114, 117, 118, 120, 121, 122, 123, 125, 126, 127, 128, 130, 133, 142, 144, 145 上, 147, 148, 149, 150, 152, 155, 156, 158, 159, 160, 163, 164, 166, 167, 168, 169, 170, 172, 173, 174, 182, 184, 186, 187, 188, 190, 191, 192, 193, 194, 196, 197, 200, 201, 202, 204, 207, 208, 209, 210, 211, 212, 213, 219, 220, 228, 229, 248, 249, 251, 252, 258, 260, 261, 262, 264, 265, 267, 268, 270, 271, 274, 276, 277, 280, 282, 283, 285, 286, 287, 289, 291, 298, 299, 300, 303, 305, 306, 308, 323, 封底

存在摄影 | Arch-Exist Photograph
226, 230, 232, 236, 238, 240, 244, 247

张东 | Dong Zhang
封面, 2, 38

赵桦 | Hua Zhao
215, 216, 217, 218, 279

姚瑜 | Yu Yao
20, 145 下, 206

张玫芳 | Meifang Zhang
215

方正 | Zheng Fang
250

宁波万科 | Ningbo Vanke
64

书中其他未列出的图片、图纸、绘制等均属于张唐景观
Credits of other unlisted photos and drawings in the book belong to Z+T Studio

图书在版编目（ＣＩＰ）数据

静谧与欢悦 = TRANQUILITY&JOY：张唐景观 Z+T STUDIO：2009—2018 / 上海张唐景观设计事务所著. —— 上海：同济大学出版社，2018.12（2020.10 重印）
 ISBN 978-7-5608-8172-0

Ⅰ. ①静⋯ Ⅱ. ①上⋯ Ⅲ. ①景观设计-作品集-中国-现代 Ⅳ. ① TU983

中国版本图书馆 CIP 数据核字 (2018) 第 220204 号

静谧与欢悦——张唐景观 Z+T STUDIO：2009 — 2018

著者：上海张唐景观设计事务所
出品人：华春荣
责任编辑：孙彬
装帧设计：姚瑜 王琪
责任校对：徐春莲

出版发行：同济大学出版社
地址：上海市杨浦区四平路 1239 号
电话：021-65985622
邮政编码：200092
网址：www.tongjipress.com.cn
经销：全国各地新华书店

印刷：上海雅昌艺术印刷有限公司
开本：787 mm X 1 092 mm　1/12
字数：680 000
印张：27
版次：2018 年 12 月第 1 版　2020 年 10 月第 3 次印刷
书号：ISBN 978-7-5608-8172-0
定价：298.00 元

版权所有 侵权必究